CELL BIOLOGY RESEARCH PROGRESS

# PROGRAMMED CELL DEATH IN PLANTS AND ANIMALS

# CELL BIOLOGY RESEARCH PROGRESS

Additional books in this series can be found on Nova's website
under the Series tab.

Additional e-books in this series can be found on Nova's website
under the e-book tab.

CELL BIOLOGY RESEARCH PROGRESS

# PROGRAMMED CELL DEATH IN PLANTS AND ANIMALS

JOSEPHINE RICE
EDITOR

nova publishers
New York

## NOTICE TO THE READER



Additional color graphics may be available in the e-book version of this book.

**Library of Congress Cataloging-in-Publication Data**
Names: Rice, Josephine, editor.
Title: Programmed cell death in plants and animals / editor, Josephine Rice.
Description: Hauppauge, New York : Nova Science Publishers, 2016. | Series: Cell biology research progress | Includes index.
Identifiers: LCCN 2015050191 (print) | LCCN 2016001135 (ebook) | ISBN 9781634845052 (hardcover) | ISBN 9781634845380 ()
Subjects: LCSH: Apoptosis.
Classification: LCC QH671 .P769 2016 (print) | LCC QH671 (ebook) | DDC 571.9/36--dc23
LC record available at http://lccn.loc.gov/2015050191

*Published by Nova Science Publishers, Inc. † New York*

# Contents

# PREFACE

Both autophagy and programmed cell death (PCD) are fundamental processes of cellular maintenance that are closely interrelated in plant and animal cells under physiological and stressful conditions. Differentiation, immune response, lack of nutrients, and wide range of abiotic factors induce their development and realization of survival or cell death scenario. Microtubular cytoskeleton is known as one of the principle players in the mediation of PCD/autophagic signals. Chapter One in this book presents an overview of the current knowledge about the role of MTs in PCD- and autophagy-related processes in plants. Chapter Two reviews the mechanisms and consequences of virus interactions with the hotel cell-death machinery, to help understand potentially pathologically relevant consequences that will help in the design of intervention strategies and the development of antiviral therapies. The final chapter discusses the control of the levels of different reactive oxygen species (ROS) and their interaction with hormones and transcriptional factors in relation with programmed cell death in leaves.

Chapter 1 – Both autophagy and programmed cell death (PCD) are fundamental processes of cellular maintenance that are closely interrelated in plant and animal cells under physiological and stressful conditions. Differentiation, immune response, lack of nutrients, and wide range of abiotic factors induce their development and realization of survival or cell death scenario. Microtubular cytoskeleton is known as one of the principle players in the mediation of PCD/autophagic signals. Microtubules (MTs) were shown involved in numeral PCD-like processes as well as in autophagosome maturation and traffic. As very dynamic structures MTs are highly sensitive to external stimuli and their functional state is basic for their role in realization of pro-death or pro-recycling cellular program. Generally, it provided with cytoskeletal network reorganization via balancing between stable/dynamic MTs behavior that is closely depends on variety of post-translational modifications of tubulin and its situational protein microenvironments. This chapter represents an attempt to overview the current knowledge about the role of MTs in PCD- and autophagy-related processes in plants.

Chapter 2 – In order to thwart a viral infection, the host has developed an integrated defense network that includes innate and adaptive immunity as well as regulated cell death (RCD), which appears to represent an ancestral defence mechanism for the host. Indeed, RCD is important to control several aspects of the cell fate in various situations, from embryonic development and adult tissue homoeostasis development, to stress response, including viral infection, where different but complementary highly regulated arms of the complex RCD process exist and can be activated, with various consequences. In order for a host to cope with a viral infection, RCD can play a critical role in the elimination of virus-infected cells and/or to control the cell-to-cell viral dissemination to protect the whole organism. During the pathogen-host co-evolution, viruses have developed various strategies to counteract defenses of the host and, among other things, they have acquired the capacity to subvert host-cell RCD. For some viruses, disabling host-cell RCD, at least for some time, could represent an important step in the replication cycle as an effort to prolong their own survival and possibly establish a persistent infection, which can lead to chronic diseases. On the other hand, other viruses may take advantage of RCD to facilitate shedding and disseminate more efficiently in the host or to prevent the presentation of viral antigens to the immune cells. By encoding numerous proteins that either are homologs to cellular apoptosis-regulatory factors or interfere with cellular proteins that control RCD-related pathways, several different viruses are able to induce or modulate the different cellular pathways that control RCD. Virus modulation of cell death participates in the pathological process associated with several human diseases; therefore understanding mechanisms and consequences of virus interactions with the host cell-death machinery is essential to understand potentially pathologically relevant consequences that will help in the design of intervention strategies and the development of antiviral therapies.

Chapter 3 – The leaf is a complex organ composed of different tissues and cells that is subjected to a variety of environmental factors affecting the course of its aging, senescence and death. Most investigations on leaf death have been focused on the senescence of photosynthetic mesophyll cells, whose characteristic loss of chlorophyll becomes the reference symptom preceding leaf death. Autumnal and monocarpic leaf deaths are developmental process genetically programmed but, at the cell level, many changes that precede them are similar to those of the senescence of shaded basal leaves and even to those of the stress-dependent leaf aging. Most knowledge of the molecular processes in programmed leaf death has been gained through the investigation with the last two deaths, provoked by developmental and environmental factors that in strict sense are not programmed leaf death. However, analogies with processes in programmed cell death in animals provide unifying views where reactive oxygen species (ROS, mainly singlet oxygen $^1O_2$, anion radical superoxide $O_2\bullet^-$ and hydrogen peroxide $H_2O_2$) emerge as key intracellular signals. The steady state level of each ROS depends on generation and scavenging activities closely linked, respectively, to the photosynthetic electron transport and to systems that protect cell from the destructive action of ROS and, frequently, the destructive actions of ROS are hardly distinguishable from their signaling role in cell death. By controlling the level of different ROS, chloroplasts regulate leaf death and assume similar roles to several ones attributed to mitochondria in programmed animal cells. High ratios ($^1O_2$ plus $O_2\bullet^-$)/$H_2O_2$ seem key to shot the autocatalytic cell path to death where hormones and transcriptional factors joint to change the feed-back regulation of cell processes to a feed-forward spiral of deleterious reactions. In this chapter, the authors will discuss the control of the levels of different ROS and their interaction with hormones and transcriptional factors in relation with programmed cell death in leaves.

In: Programmed Cell Death in Plants and Animals ISBN: 978-1-63484-505-2
Editor: Josephine Rice 

*Chapter 1*

# MICROTUBULAR CYTOSKELETON IN AUTOPHAGY AND PROGRAMMED CELL DEATH DEVELOPMENT IN PLANTS

***D. I. Lytvyn** *and Ya. B. Blume***
Department of Genomics and Molecular Biotechnology,
Institute of Food Biotechnology and Genomics,
National Academy of Sciences of Ukraine, Kyiv, Ukraine

## ABSTRACT

Both autophagy and programmed cell death (PCD) are fundamental processes of cellular maintenance that are closely interrelated in plant and animal cells under physiological and stressful conditions. Differentiation, immune response, lack of nutrients, and wide range of abiotic factors induce their development and realization of survival or cell death scenario. Microtubular cytoskeleton is known as one of the principle players in the mediation of PCD/autophagic signals. Microtubules (MTs) were shown involved in numeral PCD-like processes as well as in autophagosome maturation and traffic. As very dynamic structures MTs are highly sensitive to external stimuli and their functional state is basic for their role in realization of pro-death or pro-recycling cellular program. Generally, it provided with cytoskeletal network reorganization via balancing between stable/dynamic MTs behavior that is closely depends

* E-mail address: dmytro.lytvyn@gmail.com.

on variety of post-translational modifications of tubulin and its situational protein microenvironments. This chapter represents an attempt to overview the current knowledge about the role of MTs in PCD- and autophagy-related processes in plants.

## INTRODUCTION

Processes associated with programmed cell death (PCD) and autophagy represent an example of extremely conservative intracellular mechanisms that provide wide range of functions covering a variety of cellular events from development to survival. From an evolutionary point of view, the presence of many common features of PCD and autophagy in all eukaryotic kingdoms points their genealogical antiquity as well as significant importance for the functioning of eukaryotic cells. Plants and animals use similar cell death regulation pathways such as stress-mediated reactive oxygen species (ROS) production and the release of cytochrome *c* that suggests an existence of common unicellular ancestor of both eukaryotic kingdoms as well as antiquity of PCD-related processes [1]. Plant PCD development was described under normal physiological state and processes, including the tissue and embryo development, senescence, pollen self-incompatibility and others as well as under pathogen influence and abiotic stresses [2, 3]. Also, autophagic protein and organelle degradation/recycling represents fundamental mechanism of cellular maintenance and survival but it is also closely integrated in PCD development under varied conditions [2, 4, 5]. Microtubules (MTs) as important cellular structure are involved deeply in regulation of numerous cellular processes, such as cell division, growth, differentiation, organelle transport and cellular signaling; and, given the dynamic of nature MT, they are highly sensitive intracellular sensor for many external stimuli [6-9]. Also one of the MT functions which lacks complexity, is the involvement of MTs in the implementation of PCD and autophagy. In particular, in plants MTs are involved in such PCD-like and related processes as development, tracheary elements differentiation, leaves senescence, self-incompatibility and hypersensitive response (HR) [10]. On animal models it was shown the multiple role of MTs in autophagosome maturation and traffic [11, 12]. In the case of plants, the number of evidences that highlighting relationships between PCD and MTs is much smaller and this chapter is intended to summarize recent data in this area.

## PROGRAMMED CELL DEATH-RELATED PROCESSES

Processes related to genetically controlled and energy dependent cellular degradation, notably PCD, occur under the range of developmental events (seed formation and germination, embryo growth, pollen germination and development, xylogenesis, tissue-specific senescence, etc. [2]) as well as under external stresses including different abiotic factors and pathogenic invasions leading to hypersensitive response development [4]. To date, the majority of attempts to classify plant PCD types is based rather on the morphological changes occurring in the cell then on the clusterization of specific biochemical pathways. Animal cell death traditionally has been subdivided on three types: apoptosis, autophagy and necrosis [3]. The term "apoptosis" is commonly used for describing PCD in animals and implies caspase-3-dependent chromatin condensation and nucleosomal DNA fragmentation [13]; cytoplasm shrinkage formation or apoptotic volume decrease due to changes in $K^+/Na^+/Cl^-$ intracellular movement [14]; and fragmentation of the cell by apoptotic bodies formation, that are main morphologically identifiable features of this process.

Mentioned cellular events were described also in plant species under different experimental conditions. Thus, nucleosomal DNA fragmentation and caspase 3/7-like activity was shown as correlated with nuclear translocation of NAC089 membrane-associated transcription factor under endoplasmic reticulum-induced PCD in *Arabidopsis thaliana* [15]. Also, caspase-3-like activity was detected under aluminum-induced PCD where it was a consequences of mitochondrial transmembrane potential disrupt and cytochrome *c* realize [16]; as well as under fusaric acid (wilting agent from *Fusarium heterosporium*)-induced PCD that was accompanied with the shrinkage and vacuolization of cytoplasm, DNA fragmentation and, that is remarkable, accumulation of nitric oxide [17]. Using induced gene silencing technique, another research team revealed a mitochondrial membrane associated hexokinase 1 as a significant PCD trigger. Its inhibition also results in cytochrome *c* release, increasing of caspase-3 and -9-like activity and DNA fragmentation in *Nicotiana benthamiana* [18].

Typically, mentioned above PCD hallmarks including cell shrinkage and nuclear degradation were observed in BY-2 tobacco cells under cadmium treatment, ultraviolet B overexposure and after action of pathogen metabolites [19, 20, 21]. Additionally, another line of evidences proposes the role of phospholipase D as an important regulator of caspase-like-dependent PCD that was characterized by cell shrinkage and was possibly mediated by its product phosphatidic acid, $H_2O_2$ and ethylene [22]. Recently, subtilisin-like protease with caspase-3-like activity, named StSBTc-3, was discovered and shown to be able to induce cell shrinkage and death under pathogen attack in potato leaf cells [23]. Though, the final of this degradation process involves lysosomal decomposition of cellular material by neighbor cells of by phagocytes. The absence of this final degradation step makes the term "apoptotic-like" cell death more appropriate for plants. It seems that the most reasonable way to divide types of plant PCD is their generalizing into "autolytic" and "non-autolytic" PCD [5] or into "vacoular" and "necrotic" cell death in the similar way [24].

Autolytic PCD is characterized by the release of hydrolases from the vacuole, after vacuolar collapse, resulting in rapid clearance of the cytoplasm; chromatin condensation; and increase of vacuolar volume. Non-autolytic cell death is not always accompanied with the destruction of vacuole, but is characterized by swelling of organelles, absence of rapid cytoplasm clearance and increasing of vacuolar volume. This death type may be related to HR or necrotrophic PCD [5]. The current opinion is that mitochondria take a central place in the mediation of PCD signals via apoptotic proteins release from mitochondrial inner space [4]. In this connection, the difference between animals is that neither dividing nor decrease in size of mitochondria occuring in plants under PCD development [25]. In animals, permeabilization of mitochondrial membrane, cytochrome *c* release and further caspase induction are under Bax/Bcl-2 proteins control, but plant genomes lack the core PCD regulators such as caspases and Bcl-related proteins [4, 26]. Bax inhibitor-1 (BI-1) is one well characterized PCD suppressor discovered in plants that blocks Bax-induced cell death downstream of Bax action at the mitochondria [26, 27]. BI-1-mediated cell death regulation is seems to be closely related to endoplasmatic reticulum-stress development and its modulation of $Ca^{2+}$ signaling and sphingolipid metabolism in plant cell under different types of harmful external impacts [26].

As it was widely described, plant proteases with PCD-related functioning "metacaspases" that demonstrates caspase-like activity have no homologous in plant genomes. To date, they are divided into two sub-types: type I is characterized by N-terminal prodomain containing a proline-rich repeat motif and a zinc-finger motif; type II metacaspases lack such a prodomain but harbor a linker region between the putative large and small subunits [28]. Regulation of PCD is one of "metacaspases" functions that was well described under development and stress conditions (for review see [29-32]).

It was noted above already, that in plant cell the autophagy is an abundant precursor of cell death. Thus, many examples of developmental cell death

PCD can be referred to autophagic PCD, namely xylem development and PCD in embryo-suspensor cells. But autophagy may play a dual role as pro-survival as well as pro-death cellular mechanism [33]. Under the development of HR, autophagy exhibits pro-survival function that prevents death signal spreading between neighboring cells, and beclin, PI3K III/VPS30, ATG3 and ATG7 are involved in this process [34]. In *Arabidopsis atg6* mutant the deregulation of cell death was shown and spreading of bacterial invasion into uninfected tissue under pathogen attack. It gives the reasons to consider Atg6 (beclin1) as a significant element in autophagic regulation of HR-related cell death [35]. Also, it was proposed a pro-survival cellular scenario realizing by salicylic acid activated autophagy via NPR1 (Non-expressor of Pathogenesis-related Genes 1) and amplification of salicylic acid signaling by EDS1 (Enhanced Disease Susceptibility 1 protein)/PAD4 (Lipase-like Phytoalexin Deficient 4 Protein) during aging or pathogen infection [36, 37]. On the other hand, senescence or infection-induced accumulation of ubiquitinated proteins and ER stress are accelerated by NPR1 in *atg2* and *atg5*-deficient mutants [38]. Activity of PAD4-dependent, but not NPR1-related salicylic acid pathway also takes place in *Arabidopsis* senescence-associated ubiquitin ligase1 (SAUL1) mutant under leaf senescence cell death and the importance of SAUL1 was shown under suppression of pro-death events [39].

Recently, metacaspase-1 (AtMC1) was described as dual pro-death/pro-survival protease which cellular function is situational-dependent. In young plants, AtMC1 contributes to HR development after pathogen invasion and recognition; in this case, mentioned pro-death function of autophagy probably could be mediated by an active overload of the vacuole. Thus, in aging cells, the pro-survival functions of AtMC1 and autophagy were revealed by the constant increase of damaged proteins and organelles that accumulate in the cell and require clearance [40]. Reactive oxygen species, namely hydrogen peroxide, and nitric oxide are the most permissive pro-death factors are capable to spread death signal between cells. Vacuolar-processing enzyme (VPE) exhibits caspase-1 like activity, AvrPto-dependent Pto-interacting protein 3 and ceramides were referred among cellular mediators that are able to provide crosstalk between PCD and autophagy in plant cell [33]. For actual detailed overview of the balancing between survival and death under autophagy development see [41].

## AUTOPHAGY IN PLANT CELL

Autophagy is the bulk and universal process of macromolecular degradation and cellular remodeling of cytoplasmic materials. It realize by forming of specialized lytic membrane compartments and determines an extremely wide range of recycling processes related both to physiological cellular maintenance and to cellular answers to external adverse impacts. In plants, two major autophagic pathways have been described, microautophagy and macroautophagy [42, 43]. As reviewed in [42, 44], processes referred as microautophagy or microautophagy-like (an engulfment decomposition by vacuolar hydrolases) were described for deposition of storage proteins in developing wheat seeds, release of rubber particles into the vacuole, degradation of starch granules and storage protein in vacuoles of germinating seeds, petal senescence, during PCD in nucellus and nectar cells, and suggested as a principal process by vacuole formation in mesophyll cells of *Dendrobium petals* [45].

Macroautophagy, commonly referred as autophagy, involves the selective formation of specific double-membrane structures called autophagosomes around the cytoplasmic material to be degraded and their direct delivery for fusion with vacuole. Then, limiting with inner autophagosome membrane autophagic bodies undergo degradation by vacuolar hydrolytic enzymes. The genetic core of autophagy regulation is near completely conservative among eukaryotic kingdoms containing several discrete functional groups. First group contains genes related to autophagosome nucleation and formation of preautophagosomal structure (PAS), namely genes of Atg1 kinase complex (induction and regulation of autophagy by target of rapamycin (TOR) kinase complex), phosphatidylinositol 3-phosphate kinase (PI3K) III complex (attraction of phosphatidylinositol 3-phosphate binding proteins of Atg9 complex to the preautophagosomal structure), complex of transmembrane Atg9 proteins (provides source of the lipid membrane by PAS formation). The second group composes from genes of participating in autophagosome maturation step: Atg8-lipidation and Atg12-conjugation systems that are chargeable for autophagosomal membrane elongation (see [42, 46-48] for detailed review).

It is confidently predicted that TOR signaling pathway is the regulatory trigger of autophagy in plants [49-52]. Inhibition of yeast TORC1 (TOR complex 1, one of two multisubunit complex of TOR kinase described in yeasts and mammals [53]) by rapamycin depresses autophagy development [54] through prevention of Atg1-Atg13 complex formation [55]. Mentioned

Atg1, Atg13, TORC1 (but not TORC2) as well as homologues of yeast TORC, KOG1 and LST8 genes were found in plants [49, 52]. More further experimental proofs of TOR regulation of autophagy are required in plants, but insensitivity of many plant TOR homologues to rapamycin [52] complicates that.

Macroautophagy and macroautophagy-related processes cover a wide range of cellular events of both physiological conditions and as the reaction to adverse external factors. Indeed, in plant cells autophagy occurs not only under stressful conditions but also as a common process of cellular maintenance to provide normal physiological turnover of monomers and elimination of degraded compartments [43, 48, 56, 57]. Overall, the role of autophagy in the vacuolar biogenesis is generally accepted [48, 50], but it is suggested that under normal conditions autophagic pathways related to vacuole formation are distinct from stress-induced ones [58]. Also, autophagy was shown as involved in formation of protein storage vacuole (PSV) [59]. Autophagy was suggested as a mechanism of formation of prolamins containing PSV in maize aleurone cells, but mentioned process was considered as atypical autophagy because the lack of Atg8 into the formed membrane structures [60]. Toyooka with co-workers revealed two distinct autophagy-related processes in the cotyledon cells of germinating *Vigna mungo* seeds: starch granules wrapped with an acidic cell compartments incorporate into the lytic vacuole that is converted from PSV and subsequently degrade; and macroautophagic degradation of mitochondria and cytoplasm into autophagosomes [61]. Autophagy also was shown as a basic degradation process for starch granules in plant leaves [62, 63].

Plants, as well as other eukaryotes, show the development of autophagic scenario as a cellular answer to the lack of nutrients. Carbohydrate- and nitrogen starvation was widely observed as autophagy inducers [64-69]. Thus, autophagy is postulated as an important managing mechanism of nitrogen and carbon remobilization under both normal and starvation conditions at the level of whole plant [68, 70-72]. Also, autophagy was described as an adoptive process under salt and osmotic stress development and is capable mainly for scavenging of oxidized proteins and of ROS level regulation [73-75]. Moreover, upregulation of *AtATG18a* was shown as accompanied with the development of mentioned stresses in *Arabidopsis* [76] and was proposed as required for autophagosome formation [77, 78].

Evidence of selectivity of autophagy to its specific target stills limited in plants. It is generally considered that organelle-specific autophagy is conserved among eukaryotes [48] and the presence of plant homologues of animal autophagic cargo receptors p62 and NBR1 supports this assertion [48, 79, 80]. ATG8 proteins are postulated as a central mediator in specific recognition of cellular target to be degraded; it is provided by the presence of ATG8-interacting motif that is a unique N-terminal extension of Atg8 serving as adaptor to mark the cytosolic cargo [81]. Recently, autophagy was revealed as important regulator of peroxisomes quality by the selective degradation of oxidized peroxisomes that prevents the accumulation of damaged peroxisomes in plant cells but specific adaptor that connects the peroxisome to the phagophore membrane is not identified [82, 83]. Selective degradation of chloroplasts and Rubisco-containing body by autophagy was described as developmental and adaptation processes [84]. Also, several evidences postulate the presence of mitophagy and its role in ROS scavenging in plant cell [75].

## Plant Microtubules

In plants MTs play an essential role in cellular growth, division, development, cell signalling, organelle transport and determination of cell shape and polarity, response to hormone signals and environmental perturbations [6-9]. MTs are forming by polymerization of globular near 55-kD monomers of α- and β-tubulins. Tubulins subunits assembled in head-to-tail fashion create protofilaments; 13 protofilaments associate laterally that results in creation of hollow cylinder with an external diameter of 25 nm. Binding with guanosine 5-triphosphate (GTP) of both α- and β-tubulins is required for MT polymerization. α-Tubulin binds irreversibly but not hydrolyze GTP and are required for transverse MT orientation. GTP bound to the β-tubulin is hydrolyzed to GDP during or shortly after MT assembly. The GTP/GDP cycle is essential for the dynamic instability of MT; hydrolysis of GTP reduces the binding affinity of tubulin to adjacent molecules and provides MT depolymerization. MTs growth from γ-tubulin ring complex (the nucleation center) that is composed of γ-tubulin complex protein 1 (GCP1) and five additional proteins GCP2–GCP6 [85].

In plants as well as in animals, diversities between microtubular structures and functions are provided with combined complex realization of: expression of tubulin isoforms (for exemple, *Arabidopsis thaliana* genome has six and nine α- and β-tubulins, correspondingly); attraction of varieties of microtubule associated proteins (MAP) (as conserved in eukaryotes as well as plant-specific [7]); and multiple post-translational modifications of tubulins. The principal difference between animal and plant MTs is the disposition of their origin. In animals, MT polymerization starts in the centrosome microtubule-organizing center. This small spherical structure contains a central pair of centrioles surrounded by the pericentriolar material and is associated with nuclear envelop. It is contrasted with plants where MT growth is initiated exclusively by specialized nucleation complexes at spatially dispersed sites [8, 86].

Phases of plant cell cycle are characterized by four different MT arrays providing control of the variety of functions related to cellular growth and division: cortical array, pre-prophase band, mitotic spindle and phragmoplast. Among mentioned arrays, only mitotic spindles serve to separate chromosomes in dividing cell, others are related with development of the plant cell wall, cell shape and growth polarity [6]. Dynamic, associated with plasma membrane MTs localize in the cellular periphery and form multidirectional cortical array (cortical MTs). Cortical MTs are unstable and demonstrate dynamic switching between polymerization and disassembly.

Plant MTs behavior is regulated by numerous MAPs: as conserved among all eukaryotes (γ-tubulin complex and augmin complexes, katanin, kinesin superfamily (10 conserved families and one plant-specific ARK (armadillo repeat-containing kinesin) family), EB1, CLASP, MOR1, MAP65, TPX2 and formins), as well as plant-specific (cellulose synthase interacting MAPs, SPR2, MAP70, AIR9, WVD2/WDL, RIP/MIDD, RUNKEL, MPB2C, SPR1, MAP18/PCaP, EDE1 and MAP190) (for detailed review see [7]). Cortical MTs are crucial cellular compartment determining direction of cell growth. It was considered that cellulose microfibrils orientation depends on the special orientation of MTs and cellulose synthase interacting1 (CSI1) protein was shown as a MAP that connects cellulose synthase complexes to MT [87]. Transport of organelles and RNA granules is another function of cortical MTs that is mediated by numeral members of kinesin motors superfamily. Preprophase band, another MT array, is structure of ring-like highly aligned MTs that forms between late G2 and prophase prior mitotic spindle and phragmoplast formation and serves to point cell division site. Then, mitotic spindle is generated to segregate duplicated chromosomes, after that it is reused in the phragmoplast. This structure provides directed traffic of vesicle containing cell wall material to form cell plate between daughter cells [6].

It was noted above that MTs are dynamically unstable structures that exist in several functional states, and stable or dynamic condition of MT population causes its functional specialization. The data of *in vitro* reconstituted microtubule dynamics with pure tubulin that was summarized by Hamada [88] enabled to separate plant MAPs as: MT stabilizing proteins – MAP65 and CLASP families; microtubule destabilizers – kinesin-13, katanin, MAP18 and MDP25; and proteins that can affect MT dynamics situational – EB1, MAP215, MOR1, MAP200 and SPR2. Tubulin post-translational modifications are another MT characteristic that has significant impact on the microtubular functions; it is realized by changing of affinity of MTs to different MAPs and modulating its protein microenvironment. In this review we will focused on tubulin acetylation that was shown as a crucial post-translational modification under development of autophagic scenario in animal cell [12, 89].

## MICROTUBULES IN CELL DEATH-RELATED PROCESSES

The multiple role of microtubular cytoskeleton in the processes related to cell death is undoubted. But there are many gaps in understanding of whole interrelation picture between MTs and PCD development. This knowledge misbalance is visible for plants especially, because the bulk data in this field was generated in yeasts and animals. Since this point, we will extrapolate recent knowledge obtained in plants to complementary animal one.

### PCD-Related Organelle Traffic, Molecular Motors, and PCD-Related Maps

First of all, the MT role in PCD mediation concerns to supporting of intracellular organelle traffic. Thus, MT-dependent organelle redistribution under apoptotic development, namely mitochondria relocalization to the Golgi-proximal microtubule organizing center, is a well described process under cell exposure to different stress-inducing stimuli [90]. Human KIF5B was shown as a motor protein which activity stipulates mitochondrial and lysosomal dispersion to cell periphery [91] and, on the other hand, perinuclear mitochondrial localization facilitates BCL-2 mediated transfer of AIF (apoptosis initiating factor), Endonuclease G and ROS from the mitochondria to the nucleus that provides following DNA degradation [90, 92]. To date, we have some evidences that of KIF3C, kinesin-5 and kinesin spindle protein or their overexpression accompany tumor development and, inversely, which inhibition leads to tumor suppression [93-95] or apoptosis of tumor cells [96, 97].

KIF5B mediated mitochondrial transport was shown as dependent of the activity of JNK-signaling pathway [98], but we have no evidence of this process in plant as yet. BLAST homology search revealed several genes in *Arabidopsis* genome that have low identity with human KIF5B: At3g10180 (27.6%), At3g44050 (26.4%), BimC-type AtKin-5 At3g45850 (25.6%) and At2g28620 (26.9%), AtKin-1 At3g63480 (42.2%) and kinesin-like protein FRA1 At5g47820 (27.1%). At same time cellular functions of mentioned genes still mainly undiscovered. One of this genes, At3g44050 was shown as cell-cycle dependent kinesin which expression is OBP1 (OBF BINDING PROTEIN 1) transcription factor dependent [99] and is altered under pollen germination and pollen tube growth in *Arabidopsis* [100]; At3g44050 as well as At2g28620 was also characterized as up-regulated during mitosis [101, 102]. Interestingly, that At3g44050, At3g45850 and At2g28620 are down-regulated in the seedling-lethal plastid ATP synthase 1 mutant that is characterized by destabilized ATP synthase complex and impaired structure of the chloroplasts [103]. AtKin-1 was suggested as essential member of nuclear division cycles during gametogenesis and regulation of meiotic crossover distribution [104, 105]. Kinesin-4 FRA1 (discovered in *Arabidopsis fragile fiber* (*fra1*) mutant) was described as the motor implicated in cellulose microfibril deposition and orientation along cortical MT network by transporting Golgi-derived vesicles [106, 107] and its cellular functions is not complementary to animal homologues [108].

To date, several plant kinesins were shown associated with mitochondria and suggested as required for mitochondria positioning in plant cell. Thus, AtKP1 is normally associated with mitochondrial membrane and possibly regulate mitochondrial positioning and its respiratory [109-111]. Two mitochondria targeted kinesin-related proteins MKRP1 and MKRP2 contain an N-terminal mitochondrial targeting signal (MTS) that may play a role in plant mitochondrial dynamics [112]. It is interesting, that one of them MKRP2 (At4g39050) was revealed as required for maximal pathogen colonization by generated plant-pathogen immune system protein interaction network using effector proteins from *Pseudomonas syringae* and *Hyaloperonospora arabidopsidis* [113]. It allows speculating about possible involvement of this kinesin-related protein in the modulation of microbe-associated molecular patterns -triggered immunity and HR-related cell death. Thereby, it is far from the clear picture of the role of molecular motor proteins in PCD-related organelle (notably mitochondria) traffic.

In human cells, MAPs also was described as a factor involved in localization of mitochondria in perinuclear area. Thus, in neuronal cell, tau phosphorylation by cyclin-dependent kinase 5 (CDK5) lead to it"s realize from MTs and nuclear clustering of mitochondria. It reveals the role of tau protein in the mediation of CDK-dependent perinuclear calcium flow from endoplasmic reticulum to mitochondria in realization of neuronal cell death [114]. Tau dysfunctions and fiber abnormalities, also called "tauopathies," are connected with a list of neurodegenerative diseases. Tau action provides cell polarity and axonal outgrowth during development, but abnormal tau action can induce neuronal cell death that is caused by the inability of affected cells to properly regulate the dynamic behavior of their MTs [115]. There are two models suggesting the mechanism of tau-mediated neuronal cell death: environmental or genetic causes that induce aberrant interaction of tau with MTs and then forming aggregates and/or accumulation of cytotoxic neurofibrillary tangles in the cell [115].

In this relation, plant-related data are limited by some promising correlation between cell death processes and upregulation of MAP homologues. Previously, we have identified genes containing motifs specific to only the structural MAPs that were referred to MAP2-tau family; namely ubiquitin-associated/translation elongation factor *EF1B* At2g26920 and YELLOW-LEAF-SPECIFIC GENE 8 (*YLS8*) At5g08290 with 78.9 and 100% of correspondence to MAP, respectively [116]. Thus, *Arabidopsis* leaves senescence induced by 6-day dark treatment was characterized by well defined upregulation of *EF1B* [117]; and *YLS8* was shown as senescence-up-regulated gene which transcripts were accumulated most significantly in senescent leaves during natural senescence, but not under dark-induced artificial senescence, and under ethylene or abscisic acid treatment [118]. It gives us a possibility to suggest the role of these genes in the realization of senescence-related cell death, but relation of plant tau homologues to organelle traffic mentioned above requires further investigations.

## MTs Behavior in Plant Developmental and Induced PCD

In plants, changes in MT dynamics and cellular orientation accompany numeral death-related events (development of embryo and tracheary elements; senescence; self-incompatibility response; hypersensitive response) and mainly result in MTs depolymerization (summarized in [10]). Xylogenesis is widely described example that illustrates the interconnection between MTs and PCD under physiological developmental conditions. Rearrangement of cortical MTs network from longitudinal to lateral orientations is principle developmental point of tracheary element formation that determines orientation of cellulose microfibrils and secondary cell wall thickening and development [119, 120]. Development of secondary cell wall in *Zinnia* was shown to be dependent MAP65 protein that is responsible for the parallel grouping and bundling of cortical MTs during secondary cell wall formation [121]. NAC domain transcription factors are suggested as specific for the induction of tracheary elements development [122, 123]. Under induction of tracheary elements in *Arabidopsis in vitro* system, one of them, *VND6* was shown to be expressed followed by the secondary cell wall cellulose synthases *CesA4*, *CesA7* and nucleases *RNS3* and *Bfn1* that marked PCD [124]. This development was accompanied subsequently with linear progression in secondary cellulosic wall deposition; the arrest of secondary cellulose formation; complete nuclear degradation; shrinkage of the protoplast and concomitant cell wall lignification. Under development of mentioned events only one *AtMAP70-5* (At4g17220) was identified as specifically upregulated under tracheary element differentiation and co-regulated with mentioned above cellulose synthases and PCD-related nucleases. *AtMAP70-5* was demonstrated to bind to constitutively expressed *AtMAP70-1* (At1g68060) and these MAPs are both essential for the proper development of xylem tissue [124].

Interestingly, that MAP70-1 as well as MAP65 were shown repressed in both natural senescence and dark-induced senescence [125]. On the other hand, bundling of cortical MTs that is mediated by MAPs (especially MAP65) suggested as crucial of PCD initiation [10]. Thus, MAP65 was discovered as bound with organized MTs and their dissociation from MTs was correlated with PCD development [126]. Moreover, *Arabidopsis* leaf senescence (as natural as well as dark-induced) was accompanied by MTs disassembly, down-regulation of *MAP65, MAP70-1*, and induction of MAP18 that provides MT disassembly [125]. Alterations in MT organization also observed by PCD development in incompatible pollen under self-incompatibility induction as well as in infected cells undergoing HR. But evidences about the role of MTs in the mediation of mentioned cell death types are still limited (for more detailed information see [10]).

## MTs and Autophagy Development

Another aspect of the involvement of MT in cell death is their role in the mediation of autophagy development. It was generally accepted that in animals MTs play a dual function in autophagy: mediation of autophagosome biogenesis and its MT-dependent movement [11, 12, 127-130]. Thus, mature autophagosomes were shown as associated with MTs and concentrated near the MT-organizing center; MTs disassembly affects partially autophagosome formation, and delays their directed delivery to lysosomes [127]. Investigation of autophagosome formation and fusion with lysosome using MT-targeting drugs nocodazole and vinblastine revealed that intact MTs are required for the fusion of autophagosomes with lysosomes and that MTs also facilitate autophagosome formation [130]. Minus end-directed delivery of mature autophagosomes along MTs are required for their fusion with lysosomes; and, in animal cell, dyneins are suggested as the main motor proteins providing this traffic [128, 129]. As yet, the connection mechanism between autophagosome and dynein stills unknown, it was suggested that mentioned interaction can be directly or indirectly associated with LC8 (mammalian homologue of yeast ubiquitin-like autophagic protein Atg8) [128], this idea was supported by evidence that microinjection of antibodies against LC3 impaired autophagosome rapid movement under amino acid starvation-induced autophagy in HeLa cells [128]. But it is difficult to extrapolate these data to plant where dyneins existence and functions still debatable; thus, in rice genome, (as opposed to *Arabidopsis*) four dynein heavy chain sequences were found [131]. It possible to speculate that minus-directed kinesins can serve similar traffic functions in plants (21 C-terminal motor kinesins in *Arabidopsis* genome [132, 133].

In this relation, our data obtained by similar experimental conditions (sucrose starvation-induced autophagy in BY-2 cells) revealed that autophagosomes (stained with monodansilcadaverine) are distributed commonly at the cellular periphery, but treatment with E-64 (cysteine protease inhibitor blocking autophagy) reconstitutes autophagosome localization as described above. Autophagosomes are formed at the cellular periphery and then move bidirectionally along MTs to fuse with lysosomes. This fusion can be realized in different ways (complete fusions with forming of hybrid organelle or maintaining and traffic of two separate vesicles) suggesting different routes for autophagosome trafficking [134]. The effects observed after E-64 treatment we can explain by abnormal accumulation of autophagosomes in perinuclear area by inhibition of autophagosome fusion with vacuole. Moreover, stabilization of MTs by taxol or their depolymerization by nocodazole had the same effect as E-64 treatment (unpublished data). It gives us an opportunity to assert that MTs play fundamental role in autophagosome traffic in plant cell.

On the other hand, basing on the evidences that drug-induced MTs disruption impedes autophagy development, the role of MTs in the autophagosome formation has been postulated and this role is modified by stressful conditions (see comprehensive overview of Poüs [12] and related articles for more detailed picture of this process in animals). Molecular relationships that link MTs and autophagosome biogenesis include: interaction between LC3 and MTs, and participation of MTs and MT motors (namely, kinesin-1, kinesin-2 KIF2A and kinesin-3 KIF1B) in the regulation of mTORC1 and the class III PI 3-kinase complex involved in the initiation of the autophagic response [12].

In the case of plant autophagy, it is possible to suggest some similar mechanisms based on the investigations of plant homologues of Atg8. Initially, Atg8 was discovered in *Arabidopsis* as MT-binding protein; *Arabidopsis* Atg8 is homologue of mammalian microtubule binding light chain 3 of MAP1A and B, in spite of their absence in *Arabidopsis* genome [135]. Atg8 was shown as essential for macroautophagy development under different conditions in tobacco, *Arabidopsis* and maize [136-141]. Under starvation and leaf senescence, acyl-CoA-binding protein (ACBP3) was suggested as an important regulator of ATG8 lipidation and autophagosome formation in *Arabidopsis* [142]. En masse, functioning of Atg8 phosphatidylethanolamine lipidation system as well Atg8 functional coordination with another autophagic ubiquitin-like system Atg12 seems to be similar to animal one [137, 143-145]. In mammals, Atg8 homologues comprise of eight members [146], but plant genomes contain different number of Atg8s: nine genes in *Arabidopsis*, five genes in *Oryza sativa* and single gene in *Chlamydomonas reinhardtii* [144]. A large number of ATG8s in plants is not entirely clear, but may imply that the different ATG8 isforms possess multiple nonparallel functions in the various biological processes associated with autophagy [144].

Recently, two *Arabidopsis* Atg8-interacting proteins (ATI1 and ATI2) were identified that are unique to plants, both contain Atg8-interacting motif and interact with Atg8f isoform [147]. ATI1 and ATI2 are partially associated with the endoplasmic reticulum (ER) membrane, but carbon starvation triggers their association with the surface of spherical ER-associated compartments [147]. Starvation-induced senescence is accompanied with association of these ATI1-containing compartments with senescing plastids and their directed transport to the vacuole. It was suggested that that ATI1 is required for incorporating specific plastid proteins destined for degradation into ATI1-plastid-associated bodies by its ability to interact directly with them, and interaction of ATI1 with ATG8 eventually leads to selective autophagy-dependent transport of plastids [148].

A list of autophagy-related proteins is associated with dynamic subset of animal MTs (ATG5–ATG12, beclin 1(ATG6), WIPI1 (ATG18), ULK1/UNC-51 (ATG1), p62, AMBRA1, FYCO1 – summarized in [12]) under autophagosome biogenesis and the majority of them have plant homologues, but the evidence of complementary cellular events still to be scant in plants. Significant progress in understanding of plant autophagic cargo delivery was outlined with discovering of selective cargo receptors in plants, namely Joka2 [80], a structural homologue of p62 and NBR1 in tobacco, which simultaneously bind both ubiquitin and autophagy-specific LC3/GABARAP and are significant for selective autophagic protein degradation [149, 150]. As a functional homological hybrid of p62 and NBR1, it plays a significant role in the autophagic degradation of polyubiquitinylated proteins accumulated under stressful conditions [80, 151]. Similar functions are suggested for *Arabidopsis* homologue At NBR1 (At4g24690) that allows us to assume an existence of conservative mechanism of Atg8- associated autophagic cargo traffic in plants [79].

Not also MT dynamics but also tubulin post-translational modifications have significant impact on the MT-mediated autophagic processes. In this relation tubulin acetylation at Lys40 attracts attention in the first place [11, 12]. The main proposed function of tubulin acetylation is supporting of stable state of MTs. It was also proposed as inducer of enzymatic MT severing and as regulation factor for MAP binding by inducing conformational changes in the α/β-tubulin dimer, despite of localization of this modification on the inside of the MT polymer [152]. Tubulin acetylation was shown as required for fusion of autophagosomes with lysosomes to form autolysosome [153], and, moreover, attracting of kinesin-1 to acetylated MTs activates c-Jun N-terminal kinase-1 (JNK1) that lead to subsequent phosphorylation of Bcl-2 and Bim, relize of beclin 1 and inducing of autophagosome formation [89].

Respectively, tubulin acetylation was also described in plants [154-157]. It was shown that gibberellic acid induces tubulin acetylation and MT stabilization [158]. When tubulin acetylation was investigated in maize, tissue specific character of this modification was demonstrated [159]. Also, this tubulin modification is dynamically regulated in plant organs during development as was shown in *Brassica rapa* [160] and is important for cell division and expansion [161]. Acetylation of plant tubulin is associated with stabilized and bundled MTs under chromium stress [162] and, together with detyrosination, provides MTs stability and reduces their turnover during cadmium stress [163].

At the present, no published data are available that reveal the interrelation of tubulin acetylation and plant autophagy. But data obtained by our group bring to light convincing proofs of correlation between tubulin acetylation and starvation-induced autophagy: in BY-2 tobacco cells autophagy was strongly accompanied by α-tubulin acetylation and the termination of autophagy led to the development of PCD that was characterised by the decrease in α-tubulin acetylation [164]. Moreover, *Arabidopsis* seedlings subjected to metabolic-, salt- and osmotic stresses as well as irradiation of ultraviolet B exhibited development of autophagy that was accompanied by tubulin acetylation as well as lipidation of Atg8 with phosphatidylethanolamine (in press). Altogether, this data let us a possibility to predict an existence of mechanism of MT-mediated autophagy development in plants that is closely similar to animal one at the level of microtubular functional state.

## CONCLUSION

To date, understanding of many aspects of plant PCD are based mainly on the investigations pointed at morphological events in dying cell or organism. On the other hand, often we have to make our suggestions on the assumption of analogy with processes described in animal and/or yeasts or on the resting on comparative transcriptomic data. In plants, many PCD-related cellular events such as caspase like activities, p53 like DNA damage response activity [165] and specific organelle traffic under cell death development do exist in spite of the absence of appropriate homologues in plant genomes. It indicates the evolutional molecular divergence of crucial cellular processes that is the cause of appearance of complementary PCD mechanisms in eukaryotic kingdoms. To understand deeply MTs involvement and impacts on the different death processes at the molecular level it is principal to discover stressful regulation of microtubular functional state, namely the role of post-translational modifications of tubulins and their MAP microenvironments. Also, whereas a lot of differences between plant and animal molecular motors, it is important to specify and determine functional role of plant MAP proteins under cell death-related organelle traffic. These questions provide a wide field of further investigation of plant PCD.

## REFERENCES

[1] Blackstone, Neil. W. and Kirkwood, Thomas. B. L. 2003. "Mitochondria and Programmed Cell Death "Slave Revolt" or Community Homeostasis?" in *Genetic and Cultural Evolution of Cooperation*, edited by Peter Hammerstein, 305-25, Cambridge, MA, USA: The MIT Press.

[2] Van Doorn, Wouter G. and Woltering, Ernst J. 2005. "Many Ways to Exit? Cell Death Categories in Plants." *Trends in Plant Science* 10(3): 117–22.

[3] Mccabe, Paul F; Reape, Theresa J.; and Mccabe, Paul F. 2008. "Apoptotic-like Programmed Cell Death in Plants." *New Phytologist* 180(1): 13–26.

[4] Reape, Theresa J.; Molony, Elizabeth M. and McCabe, Paul F. 2008. "Programmed Cell Death in Plants: Distinguishing Between Different Modes." *Journal of Experimental Botany* 59(3): 435–44.

[5] Van Doorn, Wouter G. 2011. "Classes of Programmed Cell Death in Plants, Compared to Those in Animals." *Journal of Experimental Botany* 62(14): 4749–61.

[6] Wasteneys, Geoffrey O. 2002. "Microtubule Organization in the Green Kingdom: Chaos or Self-order?" *Journal of Cell Science* 115(7): 1345–54.

[7] Hamada, Takahiro. 2014. "Microtubule Organization and Microtubule-Associated Proteins in Plant Cells." *International Review of Cell and Molecular Biology* 312: 1-52.

[8] De Keijzer, Jeroen; Mulder, Bela M. and Janson, Marcel E. 2014. "Microtubule Networks for Plant Cell Division." *Systems and Synthetic Biology* 8: 187–94.

[9] Nick, Peter. 2013. "Microtubules, Signalling and Abiotic Stress." *Plant Journal* 75(2): 309–23.

[10] Smertenko, Andrei and Franklin-Tong, Vernonica, E. 2011. "Organisation and Regulation of the Cytoskeleton in Plant Programmed Cell Death." *Cell Death and Differentiation*, 18(8): 1263–70.

[11] Monastyrska, Iryna; Rieter, Ester; Klionsky, Daniel and Reggiori, Fulvio. 2010. "Multiple Roles of the Cytoskeleton in Autophagy." *Biological Reviews of the Cambridge Philosophical Society* 84(3): 431–48.

[12] Mackeh, Rafah; Perdiz, Daniel; Lorin, Séverine; Codogno, Patrice and Poüs, Christian. 2013. "Autophagy and Microtubules - New Story, Old Players." *Journal of Cell Science* 126(5): 1071–80.

[13] Porter, Alan, G. and Jänicke, Reiner, U. 1999. "Emerging Roles of Caspase-3 in Apoptosis." *Cell Death and Differentiation* 6(2): 99–104.

[14] Bortner, Carl D. and Cidlowski, John A. 2007. "Cell Shrinkage and Monovalent Cation fluxes: Role in Apoptosis." *Archives of Biochemistry and Biophysics* 462(2): 176–88.

[15] Yang, Zheng-Ting; Wang, Mei-Jing; Sun, Ling; Lu, Sun-Jie; Dong-Ling, Bi; Le, Sun; Ze-Ting, Song; Shuang-Shuang, Zhang; Shun-Fan, Zhou; Jian-Xiang, Liu. 2014. "The Membrane-Associated Transcription Factor NAC089 Controls ER-Stress-Induced Programmed Cell Death in Plants." *PLoS Genetics* 10(3): e1004243. DOI: 10.1371/journal.pgen.1004243.

[16] Li, Zhe and Xing, Da. 2010 "Mitochondrial Pathway Leading to Programmed Cell Death Induced by Aluminum Phytotoxicity in *Arabidopsis*." *Plant Signaling & Behavior* 5(12): 1660–62.

[17] Jiao, Jiao; Zhou, Benguo; Zhu, Xiaoping; Gao, Zhengliang and Liang, Yuancun. 2013. "Fusaric Acid Induction of Programmed Cell Death Modulated Through Nitric Oxide Signalling in Tobacco Suspension Cells." *Planta* 238(4): 727–37.

[18] Kim, Moonil; Lim, Jeong-Hwa; Ahn, Chang Sook; Park, Kyoungsook; Gyung, Tae, Kim; Woo, Taek, Kim; Hyun-Sook, Pai. 2006. "Mitochondria-associated Hexokinases Play a Role in the Control of Programmed Cell Death in *Nicotiana benthamiana*." *Plant Cell* 18(9): 2341–55.

[19] Kutik, Jaromir; Kuthanova, Andrea; Smertenko, Andrei; Fischer, Lukas; Opatrny, Zdenek. 2014. "Cadmium-induced Cell Death in BY-2 Cell Culture Starts with Vacuolization of Cytoplasm and Terminates with Necrosis." *Physiologia Plantarum* 151(4): 423–33.

[20] Lytvyn, Dmytro I.; Yemets, Alla I.; Blume, Yaroslav B. 2010. "UV-B Overexposure Induces Programmed Cell Death in a BY-2 Tobacco Cell Line." *Environmental and Experimental Botany* 68(1): 51–7.

[21] Cheng, Dan-Dan; Jia, Yu-Jiao; Gao, Hui-Yuan; Zhang, Li-Tao; Zi-Shan; Zhang; Zhong-Cai, Xue; Qing-Wei, Meng. (2011) "Characterization of the Programmed Cell Death Induced by Metabolic Products of *Alternaria alternata* in Tobacco BY-2 Cells." *Physiologia Plantarum* 141(2): 117–29.

[22] Iakimova, Elena T.; Michaeli, Rina and Woltering, Ernst J. 2013. "Involvement of Phospholipase D-related Signal Transduction In chemical-induced Programmed Cell Death in Tomato Cell Cultures." *Protoplasma* 250(5): 1169–83.

[23] Fernandez, María Belen; Daleo, Gustavo Raúl and Guevara, María Gabriela. 2014. "Isolation and Characterization of a *Solanum tuberosum* Subtilisin-like Protein with Caspase-3 Activity (StSBTc-3)." *Plant Physiology and Biochemistry* 86: 137–46.

[24] Van Doorn, W. G.; Beers, E. P.; Dangl, J. L.; Franklin-Tong, V. E.; Gallois, P; Hara-Nishimura, I; Jones, A. M.; Kawai-Yamada, M.; Lam, E.; Mundy, J.; Mur, L. A.; Petersen, M.; Smertenko, A.; Taliansky, M.; Van Breusegem, F.; Wolpert, T.; Woltering, E.; Zhivotovsky, B.; Bozhkov, P. V. 2011. "Morphological Classification of Plant Cell Deaths." *Cell Death and Differentiation* 18(8): 1241–46.

[25] Scott, Iain and Logan, David C. 2008. "Mitochondria and Cell Death Pathways in Plants." *Plant Signaling & Behavior*. 3(7):475–77.

[26] Ishikawa, T.; Watanabe, N.; Nagano, M.; Kawai-Yamada, M.; Lam, E. 2011. "Bax Inhibitor-1: a Highly Conserved Endoplasmic Reticulum-Resident Cell Death Suppressor." *Cell Death and Differentiation* 18(8): 1271–78.

[27] Watanabe, Naohide and Lam, Eric. 2009. "Bax inhibitor-1, a Conserved Cell Death Suppressor, Is a Key Molecular Switch Downstream From a Variety of Biotic and Abiotic Stress Signals in Plants." *International Journal of Molecular Sciences* 10(7): 3149–67.

[28] Tsiatsiani, Liana; Van Breusegem, Frank; Gallois, Patrick; Zavialov, Anton; Lam, Eric; Bozhkov. Peter, V. 2011. "Metacaspases." *Cell Death and Differentiation* 18(8): 1279–88.

[29] Bonneau, Laurent; Ge, Yuan; Drury, Georgina E. and Gallois, Patrick. 2008. "What Happened to Plant Caspases?" *Journal of Experimental Botany* 59(3): 491–99.

[30] Minina, Elena, A.; Stael, Simon; Breusegem, Frank, Van; Bozhkov, Peter, V. 2014. "Plant Metacaspase Activation and Activity." *Methods in Molecular Biology*. 1133: 237–53.

[31] Fagundes, David; Bohn, Bianca; Cabreira, Caroline; Leipelt, Fábio; Dias, Nathalia; Bodanese-Zanettini, Maria, H.; Cagliari, Alexandro. 2015. "Caspases in Plants: Metacaspase Gene Family in Plant Stress Responses." *Functional & Integrative Genomics* Epub ahead of print. http://link.springer.com/10.1007/s10142-015-0459-7.

[32] Suarez, Maria F.; Filonova, Lada H.; Smertenko, Andrei; Savenkov, Eugene I.; Clapham, David, H.; von Arnold, Sara; Zhivotovsky, Boris; Bozhkov, Peter, V. 2004. "Metacaspase-Dependent Programmed Cell Death Is Essential for Plant Embryogenesis." *Current Biology* 14(9): 339–40.

[33] Patel, Shalaka; Caplan, Jeffrey; Dinesh-Kumar, S. P. 2006. "Autophagy in the Control of Programmed Cell Death." *Current Opinion in Plant Biology* 9(4): 391–6.

[34] Liu, Yule; Schiff, Michael; Czymmek, Kirk; Tallóczy, Zsolt; Levine, Beth; Dinesh-Kumar, S., P. 2005. "Autophagy Regulates Programmed Cell Death During the Plant Innate Immune Response." *Cell* 121(4): 567–77.

[35] Patel, S., and Dinesh-Kumar, S. P. 2008. "Arabidopsis ATG6 is Required to Limit the Pathogen-Associated Cell Death Response." *Autophagy* 4(1): 20–7.

[36] Yoshimoto, Kohki; Jikumaru, Yusuke; Kamiya, Yuji; Kusano, Miyako; Consonni, Chiara; Panstruga, Ralph; Ohsumi, Yoshinori; Shirasu, Ken. 2009. "Autophagy Negatively Regulates Cell Death by Controlling NPR1-Dependent Salicylic Acid Signaling During Senescence and the Innate Immune Response in *Arabidopsis*." *Plant Cell* 21(9): 2914–27.

[37] Yoshimoto, Kohki. 2010. "Physiological Roles of Autophagy in Plants: Does Plant Autophagy Have a Pro-Death Function?" *Plant Signaling & Behavior* 5(5): 494–6.

[38] Munch, David; Rodriguez, Eleazar; Bressendorff, Simon; Park, Ohkmae, K.; Hofius, Daniel; Petersen, Morten. 2014. "Autophagy Deficiency Leads to Accumulation of Ubiquitinated Proteins, ER Stress, and Cell Death in *Arabidopsis*." *Autophagy* 10(9): 1–9.

[39] Vogelmann, Katja; Drechsel, Gabriele; Bergler, Johannes; Subert, Christa; Philippar, Katrin; Soll, Jürgen; Engelmann, Julia, C.; Engelsdorf, Timo; Voll, Lars, M.; Hoth, Stefan. 2012 "Early Senescence and Cell Death in *Arabidopsis* Saul1 Mutants Involves the PAD4-Dependent Salicylic Acid Pathway." *Plant Physiology* 159(4): 1477–87.

[40] Coll, Núria, S.; Smidler, Andrea; Puigvert, Marina; Popa, Crina; Valls, Marc; Dangl, Jeffery, L. 2014 "The Plant Metacaspase AtMC1 in pathogen-Triggered Programmed Cell Death and Aging: Functional Linkage with Autophagy." *Cell Death and Differentiation* 21(9): 1–10.

[41] Minina, Elena A.; Bozhkov, Peter, V.; Hofius, Daniel. 2014. "Autophagy as Initiator or Executioner of Cell Death." *Trends in Plant Science* 19(11): 692–7.

[42] Bassham, Diane, C. 2007. "Plant Autophagy-More Than a Starvation Response." *Current Opinion in Plant Biology* 10(6): 587–93.

[43] Bassham, Diane, C.; Laporte, Marianne; Marty, Francis; Moriyasu, Yuji; Ohsumi, Yoshinori; Olsen, Laura, J.; Yoshimoto, Kohki. 2006.

"Autophagy in Development and Stress Responses of Plants." *Autophagy* 2(1): 2–11.

[44] Van Doorn, Wouter, G.; Papini, Alessio. 2013. "Ultrastructure of Autophagy in Plant Cells: A Review." *Autophagy* 9(12): 1922–36.

[45] Van Doorn, Wouter G.; Kirasak, Kanjana; Ketsa, Saichol. 2015. "Macroautophagy and Microautophagy in Relation to Vacuole Formation in Mesophyll Cells of Dendrobium Tepals." *Journal of Plant Physiology* 177: 67–73.

[46] Kim, Soon Hee; Kwon, Chian; Lee, Jae Hoon; Chung, Taijoon. 2012. "Genes for Plant Autophagy: Functions and Interactions." *Molecules and Cells* 34(5): 413–23.

[47] Yoshimoto, Kohki. 2012. "Beginning to Understand Autophagy, an Intracellular Self-Degradation System in Plants." *Plant and Cell Physiology* 53(8): 1355–65.

[48] Liu, Yimo, and Bassham, Diane, C. 2012 "Autophagy: Pathways for Self-Eating in Plant Cells." *Annual Review of Plant Biology* 63(1): 215–37.

[49] Díaz-Troya, Sandra; Pérez-Pérez, María Esther; Florencio, Francisco J., and Crespo, José L. 2008. "The role of TOR in Autophagy Regulation From Yeast to Plants and Mammals." *Autophagy* 4(7): 851–65.

[50] Bassham, Diane, C. 2009. "Function and Regulation of Macroautophagy in Plants." *Biochimica et Biophysica Acta - Molecular Cell Research* 1793(9): 1397–403.

[51] Liu, Yimo and Bassham, Diane, C. 2010 "TOR Is a Negative Regulator of Autophagy in *Arabidopsis thaliana*." *PloS one* 5(7): e11883. DOI: 10.1371/journal.pone.0011883.

[52] John, Florian; Roffler, Stefan; Wicker, Thomas; Ringli, Christoph. 2011. "Plant TOR Signaling Components." *Plant Signaling & Behavior* 6(11): 1700–5.

[53] Wullschleger, Stephan; Loewith, Robbie; Hall, Michael N. 2006. "TOR Signaling in Growth and Metabolism." *Cell* 124(3): 471–84.

[54] Noda, Takeshi and Ohsumi, Yoshinori. 1998. "Tor, a Phosphatidylinositol Kinase Homologue, Controls Autophagy in Yeast." *Journal of Biological Chemistry* 273(7): 3963–66.

[55] Kamada, Yoshiaki; Funakoshi, Tomoko; Shintani, Takahiro; Nagano, Kazuya; Ohsumic, Mariko; Ohsumia, Yoshinori. 2000. "Tor-Mediated Induction of Autophagy Via an Apg1 Protein Kinase Complex." *Journal of Cell Biology* 150(6): 1507–13.

[56] Yano, Kanako; Suzuki, Takao; Moriyasu, Yuji. 2007. "Constitutive Autophagy in Plant Root Cells." *Autophagy* 3(4): 360–62.

[57] Inoue, Yuko; Suzuki, Takao; Hattori, Masaki; Yoshimoto, Kohki; Ohsumi, Yoshinori; Moriyasu, Yuji. 2006 "AtATG genes, Homologs of Yeast Autophagy Genes, are Involved in Constitutive Autophagy in *Arabidopsis* Root Tip Cells." *Plant and Cell Physiology* 47(12): 1641–52.

[58] Yano, Kanako; Hattori, Masaki; Moriyasu, Yuji. 2007. "A Novel Type of Autophagy Occurs Together with Vacuole Genesis in Miniprotoplasts Prepared from Tobacco Culture Cells." *Autophagy* 3(3): 215–21.

[59] Ibl, Verena and Stoger, Eva. 2012 "The Formation, Function and Fate of protein Storage Compartments in Seeds." *Protoplasma* 249(2): 379–92.

[60] Reyes, Francisca, C.; Chung, Taijoon; Holding, David; Jung, Rudolf; Vierstra, Richard; Otegui, Marisa, S. 2011. "Delivery of Prolamins to the Protein Storage Vacuole in Maize Aleurone Cells." *Plant Cell* 23(2): 769–84.

[61] Toyooka, Kiminori; Okamoto, Takashi; Minamikawa, Takao. 2001. "Cotyledon cells of Vigna Mungo Seedlings Use at Least Two Distinct Autophagic Machineries for Degradation of Starch Granules and Cellular Components." *Journal of Cell Biology* 154(5): 973–82.

[62] Wang, Yan; Yu, Bingjie: Zhao, Jinping: Guo, Jiangbo; Li, Ying; Han, Shaojie; Huang, Lei; Du, Yumei; Hong, Yiguo; Tang, Dingzhong; Liu, Yule. 2013 "Autophagy Contributes to Leaf Starch Degradation." *Plant Cell* 25(4): 1383–99.

[63] Wang, Yan and Liu, Yule. 2013. "Autophagic Degradation of Leaf Starch in Plants." *Autophagy* 9(8): 1247–48.

[64] Moriyasu, Yuji and Ohsumi, Yoshinory. 1996. "Autophagy in Tobacco Suspension-Cultured Cells in Response to Sucrose Starvation." *Plant Physiology*, 111(4): 1233–41.

[65] Voitsekhovskaja, Olga V.; Schiermeyer, Andreas; Reumann, Sigrun. 2014. "Plant Peroxisomes Are Degraded by Starvation-Induced and Constitutive Autophagy in Tobacco BY-2 Suspension-Cultured Cells." *Frontiers in Plant Science* 5: 629. DOI: 10.3389/fpls.2014.00629.

[66] Doelling, Jed, H.; Walker, Joseph, M.; Friedman, Eric. M.; Thompson, Allison, R.; Vierstra, Richard D. 2002. "The APG8/12-Activating Enzyme APG7 Is Required for Proper Nutrient Recycling and Senescence in *Arabidopsis thaliana*." *Journal of Biological Chemistry* 277(36): 33105–14.

[67] Takatsuka, Chihiro; Inoue, Yuko; Matsuoka, Ken; Moriyasu, Yuji. 2004. "3-Methyladenine Inhibits Autophagy in Tobacco Culture Cells Under Sucrose Starvation Conditions." *Plant and Cell Physiology* 45(3): 265–74.

[68] Guiboileau, Anne; Yoshimoto, Kohki; Soulay, Fabienne; Bataillé, Marie-Paule; Avice, Jean-Christophe; Masclaux-Daubresse, Céline. (2012) "Autophagy Machinery Controls Nitrogen Remobilization at the Whole-Plant Level Under Both Limiting and Ample Nitrate Conditions in *Arabidopsis*." *New Phytologist* 194(3): 732–40.

[69] Guiboileau, Anne; Avila-Ospina, Liliana; Yoshimoto, Kohki; Soulay, Fabienne; Azzopardi, Marianne; Marmagne, Anne; Lothier, Jeremy; Masclaux-Daubresse, Celine. 2013. "Physiological and Metabolic Consequences of Autophagy Deficiency for the Management of Nitrogen and Protein Resources in *Arabidopsis* Leaves Depending on Nitrate Availability." *New Phytologist* 199(3): 683–94.

[70] Avin-Wittenberg, Tamar; Bajdzienko, Krzysztof; Wittenberg, Gal; Alseekh, Saleh; Tohge, Takayuki; Bock, Ralph; Giavalisco, Patrick; Fernie, Alisdair, R. 2015. "Global Analysis of the Role of Autophagy in Cellular Metabolism and Energy Homeostasis in *Arabidopsis* Seedlings Under Carbon Starvation." *Plant Cell* 27(2): 306-22.

[71] Ren, Chenxia; Liu, Jingfang; Gong, Qingqiu. 2014. "Functions of autophagy in Plant Carbon and Nitrogen Metabolism." *Frontiers in Plant Science* 5:301. DOI: 10.3389/fpls.2014.00301.

[72] Masclaux-Daubresse, Celine. 2014. "Autophagy Controls Carbon, Nitrogen, and Redox Homeostasis in Plants." *Autophagy* e36261. DOI: 10.4161/auto.36261.

[73] Xiong, Yan; Contento, Anthony, L; Nguyen, Phan Quang; Bassham, Diane C. 2007. "Degradation of Oxidized Proteins by Autophagy During Oxidative Stress in *Arabidopsis*." *Plant Physiology* 143(1): 291–99.

[74] Han, Shaojie; Yu, Bingjie; Wang, Yan; Liu, Yule. 2011. "Role of Plant Autophagy in Stress Response." *Protein and Cell* 2(10): 784–91.

[75] Minibayeva, Farida; Dmitrieva, Svetlana; Ponomareva, Anastasia; Ryabovol, Victoria. 2012. "Oxidative Stress-Induced Autophagy in Plants: The Role of Mitochondria." *Plant Physiology and Biochemistry* 59: 11–9.

[76] Liu, Y.; Xiong, Y. and Bassham, D.C. 2009 "Autophagy is Required for Tolerance of Drought and Salt Stress in Plants." *Autophagy* 5(7): 954–63.

[77] Xiong, Yan; Contento, Anthony, L.; Bassham, Diane, C. 2007. "Disruption of Autophagy Results in Constitutive Oxidative Stress in *Arabidopsis*." *Autophagy* 3(3): 257–8.

[78] Xiong, Yan; Contento, Anthony, L.; Bassham, Diane. C.. 2005. "AtATG18a Is Required for the formation of Autophagosomes During Nutrient Stress and Senescence in *Arabidopsis Thaliana*." *Plant Journal* 42: 535–46.

[79] Svenning, Steingrim; Lamark, Trond; Krause, Kirsten; Johansen, Terje. 2011. "Plant NBR1 Is a Selective Autophagy Substrate and a Functional Hybrid of the Mammalian Autophagic Adapters NBR1 and p62/SQSTM1." *Autophagy* 7(9): 993–1010.

[80] Zientara-Rytter, Katarzyna; Łukomska, Jolanta; Moniuszko, Grzegorz; Gwozdecki, Rafał; Surowiecki, Przemysław; Lewandowska, Małgorzata; Liszewska, Frantz; Wawrzyńska, Anna; Sirko, Agnieszka. 2011. "Identification and Functional Analysis of Joka2, a Tobacco Member of the Family of Selective Autophagy Cargo Receptors." *Autophagy* 7(10): 1145–58.

[81] Michaeli, Simon and Galili, Gad. 2014. "Degradation of Organelles or Specific Organelle Components Via Selective Autophagy in Plant Cells." *International Journal of Molecular Sciences* 15(5): 7624–38.

[82] Shibata, Michitaro; Oikawa, Kazusato; Yoshimoto, Kohki: Goto-Yamada, Shino; Mano, Shoji; Yamada, Kenji; Kondo, Maki; Hayashi, Makoto; Sakamoto, Wataru; Ohsumi, Yoshinori; Nishimura, Mikio. 2014. "Plant Autophagy Is Responsible for Peroxisomal Transition and Plays an Important Role in the Maintenance of Peroxisomal Quality." *Autophagy* 10(5): 936–37.

[83] Lee, Han Nim; Kim, Jimi; Chung, Taijoon. 2014. "Degradation of Plant Peroxisomes by Autophagy." *Frontiers in Plant Science* 5: 139. DOI: 10.3389/fpls.2014.00139.

[84] Ishida, Hiroyuki; Izumi, Masanori; Wada, Shinya; Makino, Amane. 2014. "Roles of Autophagy in Chloroplast Recycling." *Biochimica et Biophysica Acta - Bioenergetics* 1837(4): 512–21.

[85] Struk, Sylwia and Dhonukshe, Pankaj. 2014. "MAPs: Cellular Navigators for Microtubule Array Orientations in *Arabidopsis*." *Plant Cell Reports* 33(1): 1–21.

[86] Teixidó-Travesa, Neus; Roig, Joan; Lüders, Jens. 2012. "The Where, When and How of Microtubule Nucleation - One Ring to Rule Them All." *Journal of Cell Science* 125(19): 4445–56.

[87] Li, S.; Lei, L.; Somerville, C. R.; Gu, Y. 2012. "Cellulose Synthase Interactive Protein 1 (CSI1) Links Microtubules and Cellulose Synthase Complexes." *Proceedings of the National Academy of Sciences* 109(1): 185–90.

[88] Hamada, Takahiro. 2014. "Lessons From In Vitro Reconstitution Analyses of Plant Microtubule-Associated Proteins." *Frontiers in Plant Science* 5: 409. DOI: 10.3389/fpls.2014.00409.

[89] Geeraert, Camille; Ratier, Ameetha; Pfisterer, Simon, G.; Perdiz, Daniel; Cantaloube, Isabelle; Rouault, Audrey; Pattingre, Sophie; Proikas-Cezanne, Tassula; Codogno, Patrice; Poüs, Christian. 2010. "Starvation-Induced Hyperacetylation of Tubulin Is Required For the Stimulation of Autophagy By Nutrient Deprivation." *Journal of Biological Chemistry* 285(31): 24184–94.

[90] Aslan, Joseph, E. and Thomas, Gary. 2009. "Death By Committee: Organellar Trafficking and Communication in Apoptosis." *Traffic* 10(10): 1390–404.

[91] Cardoso, Carla, M. P.; Groth-Pedersen, Line; Høyer-Hansen, Maria, Kirkegaard, Thomas; Corcelle, Elizabeth; Andersen, Jens, S.; Jäättelä, Marja; Nylandsted, Jesper. 2009. "Depletion of.kinesin 5B affects Lysosomal Distribution and Stability and Induces Peri-Nuclear Accumulation of Autophagosomes in Cancer Cells." *PLoS ONE*, 4(2): e4424. DOI: 10.1371/journal.pone.0004424.

[92] Zinkel, S.; Gross, A. and Yang, E. 2006. "BCL2 Family in DNA Damage and cell Cycle Control." *Cell Death and Differentiation* 13(8): 1351–59.

[93] Wang, Chengqin; Wang, Chenggang; Wei, Zhimin; Li, Yujun; Wang, Wenhong; Li, Xia; Zhao, Jing; Zhou Xuan; Qu, Xun; Xiang, Fenggang. 2015. "Suppression of Motor Protein KIF3C Expression Inhibits Tumor Growth and Metastasis in Breast Cancer by Inhibiting TGF-β Signaling." *Cancer Letters* 368(1): 105–14.

[94] Guido, B. C.; Ramos, L. M.; Nolasco, D. O.; Nobrega, C.C.; Andrade, Bárbara, Y., G.; Pic-Taylor, Aline; Neto Brenno, A., D.; Corrêa, José, R. 2015. "Impact of Kinesin Eg5 Inhibition By 3,4-Dihydropyrimidin-2(1H)-One Derivatives On Various Breast Cancer Cell Features." *BMC Cancer* 15(1): 283. DOI: 10.1186/s12885-015-1274-1.

[95] Song, In-Sung; Jeong, Yu Jeong; Nyamaa, Bayalagmaa; Jeong, Seung Hun; Kim, Hyung Kyu; Kim, Nari; Ko, Kyung Soo; Rhee, Byoung Doo; Han, Jin. 2015 "KSP Inhibitor SB743921 Induces Death of Multiple

Myeloma Cells Via Inhibition of the NF-κB Signaling Pathway." *BMB Reports* 48(10): 571-6.

[96] Tang, Y.; Orth, J. D.; Xie, T.; Mitchison, T. J. 2011. "Rapid Induction of Apoptosis During Kinesin-5 Inhibitor-Induced Mitotic Arrest in HL60 cells." *Cancer Letters* 310(1): 15–24.

[97] Doan, Chinh Chung; Doan, Ngoc Trung; Nguyen, Quang Huy; Nguyen, Minh, Hoa; Do, Minh Si; Le, Van Dong. 2015 "Downregulation of Kinesin Spindle Protein Inhibits Proliferation, Induces Apoptosis and Increases Chemosensitivity in Hepatocellular Carcinoma Cells." *Iranian Biomedical Journal* 19(1): 1–16.

[98] Stagi, Massimiliano; Gorlovoy, Philipp; Larionov, Sergey; Takahashi, Kazuya; Neumann, Harald. 2006. "Unloading Kinesin Transported Cargoes From the Tubulin Track Via the Inflammatory c-Jun N-Terminal Kinase Pathway." *FASEB Journal* 20(14): 2573–75.

[99] Skirycz, Aleksandra; Radziejwoski, Amandine; Busch, Wolfgang; Hannah, Matthew, A.; Czeszejko, Joanna; Kwaśniewski, Mirosław; Zanor, Maria-Inès; Lohmann, Jan, U.; De Veylder, Lieven; Witt, Isabell; Mueller-Roeber, Bernd. 2008. "The DOF Transcription Factor OBP1 Is Involved in Cell Cycle Regulation in *Arabidopsis thaliana*." *Plant Journal* 56(5): 779–92.

[100] Wang, Yi; Zhang, Wen-Zheng; Song, Lian-Fen; Zou, Jun-Jie; Su, Zhen; Wu, Wei-Hua. 2008. "Transcriptome Analyses Show Changes in Gene Expression to Accompany Pollen Germination and Tube Growth in *Arabidopsis*." *Plant Physiology* 148(3): 1201–11.

[101] Vanstraelen, Marleen; Inzé, Dirk; Geelen, Danny. 2006. "Mitosis-Specific Kinesins in *Arabidopsis*." *Trends in Plant Science* 11(4): 167–75.

[102] Bannigan, Alex; Scheible, Wolf-Rüdiger; Lukowitz, Wolfgang; Fagerstrom, Carey; Wadsworth, Patricia; Somerville, Chris; Baskin, Tobias I. 2007. "A Conserved Role for Kinesin-5 in Plant Mitosis." *Journal of Cell Science* 120(16): 2819–27.

[103] Dal Bosco, Cristina; Lezhneva, Lina; Bieh, Alexander; Leister, Dario; Strotmann, Heinrich; Wanner, Gerd; Meurer, Jörg. 2004. "Inactivation of the Chloroplast ATP Synthase γ Subunit Results in High Non-photochemical Fluorescence Quenching and Altered Nuclear Gene Expression in *Arabidopsis thaliana*." *Journal of Biological Chemistry* 279(2): 1060–69.

[104] Wang, Haiqing; Liu, Ruijuan; Wang, Jianwu; Wang, Pei; Shen, Yuhu; Liu, Guoqin. 2014. "The *Arabidopsis* Kinesin Gene AtKin-1 Plays a

Role in the Nuclear Division Process During Megagametogenesis." *Plant Cell Reports* 33: 819–28.

[105] Duroc, Yann; Lemhemdi, Afef; Larcheveque, Cécile; Hurel, Aurélie; Cuacos, Maria; Cromer, Laurence; Horlow, Christine; Armstrong, Susan, J.; Chelysheva, Liudmila; Mercier, Raphael. 2014 "The Kinesin AtPSS1 Promotes Synapsis and Is Required for Proper Crossover Distribution in Meiosis." *PLoS Genetics* 10(10), p. e1004674. DOI: 10.1371/journal.pgen.1004674.

[106] Zhong, Ruiqin; Burk, David, H.; Morrison, W. Herbert; Ye, Zheng-Hua. 2002. "A Kinesin-Like Protein Is Essential for Oriented Deposition of Cellulose Microfibrils and Cell Wall Strength." *Plant Cell* 14(12): 3101–17.

[107] Zhu, Chuanmei; Ganguly, Anindya; Baskin, Tobias, I.; McClosky, Daniel. D.; Anderson, Charles, T.; Foster, Cliff; Meunier, Kristoffer, A.; Okamoto, Ruth; Berg, Howard; Dixit, Ram. 2015. "The FRA1 Kinesin Contributes to Cortical Microtubule-Mediated Trafficking of Cell Wall Components." *Plant Physiology* 167: 780–92.

[108] Kong, Zhaosheng; Ioki, Motohide; Braybrook, Siobhan; Li, Shundai; Ye, Zheng-Hua; Lee, Yuh-Ru, Julie; Hotta, Takashi; Chang, Anny; Tian, Juan; Wang, Guangda; Liu, Bo. 2015. "Kinesin-4 Functions in Vesicular Transport on Cortical Microtubules and Regulates Cell Wall Mechanics During Cell Elongation in Plants." *Molecular Plant* 8(7): 1011–23.

[109] Ni, Cheng Zhi; Wang, Hai Qing; Xu, Tao; Qu, Zhe; Liu, Guo, Qin. 2005. "AtKP1, a Kinesin-Like Protein, Mainly Localizes to Mitochondria in *Arabidopsis thaliana*." *Cell Research* 15(9): 725–33.

[110] Yang, Xue-Yong; Chen, Zi-Wei; Xu, Tao; Qu, Zhe; Pan, Xiao-Di; Qin, Xing-Hua; Ren, Dong-Tao; Liu, Guo-Qin. 2011 "*Arabidopsis* Kinesin KP1 Specifically Interacts with VDAC3, a Mitochondrial Protein, and Regulates Respiration During Seed Germination At Low Temperature." *Plant Cell* 23(3): 1093–106.

[111] Cai, Giampiero and Cresti, Mauro. 2012. "Are Kinesins Required for Organelle Trafficking in Plant Cells?" *Frontiers in Plant Science* 3: 170. DOI: 10.3389/fpls.2012.00170.

[112] Itoh, R.; Fujiwara, M. and Yoshida, S. 2001. "Kinesin-Related Proteins With a Mitochondrial Targeting Signal." *Plant Physiology* 127(3): 724–26.

[113] Mukhtar, M., Shahid; Carvunis, Anne-Ruxandra; Dreze, Matija; Epple, Petra; Steinbrenner, Jens; Moore, Jonathan; Tasan, Murat; Galli, Mary; Hao, Tong; Nishimura, Marc, T.; Pevzner, Samuel, J.; Donovan, Susan,

E.; Ghamsari, Lila; Santhanam, Balaji; Romero, Viviana; Poulin, Matthew, M.; Gebreab, Fana; Gutierrez, Bryan, J.; Tam, Stanley; Monachello, Dario; Boxem, Mike; Harbort, Christopher, J.; McDonald, Nathan; Gai, Lantian; Chen, Huaming; He, Yijian; Vandenhaute, Jean; Roth, Frederick, P.; Hill, David, E.; Ecker, Joseph, R.; Vidal, Marc; Beynon, Jim; Braun, Pascal; Dangl Jeffery, L. 2011. "Independently Evolved Virulence Effectors Converge Onto Hubs in a Plant Immune System Network." *Science* 333(6042): 596–601.

[114] Darios, Frédéric; Muriel, Marie-Paule; Khondiker, Myriam, Escobar; Brice, Alexis; Ruberg, Merle. 2005. "Neurotoxic Calcium Transfer From Endoplasmic Reticulum to Mitochondria Is Regulated By Cyclin-Dependent Kinase 5-Dependent Phosphorylation of Tau." *Journal of Neuroscience* 25(16): 4159–68.

[115] Feinstein, Stuart, C. and Wilson, Leslie. 2005. "Inability of Tau to Properly Regulate Neuronal Microtubule Dynamics: A Loss-Of-Function Mechanism By Which Tau Might Mediate Neuronal Cell Death." *Biochimica et Biophysica Acta - Molecular Basis of Disease* 1739(2): 268–79.

[116] Karpov, Pavel, A. and Blume, Yaroslav, B. 2008. "Bioinformatic Search For Plant Homologues of Animal Structural MAPs in the *Arabidopsis Thaliana* Genome." In *The Plant Cytoskeleton: Key Tool for Agro-Biotechnology*, edited by Yaroslav B. Blume, Vance W. Baird, Alla I. Yemets and Diego Breviari, 373–94: Springer Netherlands.

[117] Lin, Ji, Feng and Wu, Shu, Hsing. 2004. "Molecular Events in Senescing *Arabidopsis* Leaves." *Plant Journal* 39(4): 612–28.

[118] Yoshida, S.; Ito, M.; Nishida, I.; Watanabe, A. 2001 "Isolation and RNA gel Blot Analysis of Genes That Could Serve as Potential Molecular Markers for Leaf Senescence in *Arabidopsis Thaliana*." *Plant & Cell Physiology* 42(2): 170–78.

[119] Fukudai, Hiroo and Kobayashi, Hitoshi. 1989. "Dynamic Organization of the Cytoskeleton During Tracheary-Element Differentiation." *Development, Growth & Differentiation* 31(1): 9–16.

[120] Oda, Yoshihisa; Mimura, Tetsuro; Hasezawa, Seiichiro. 2005. "Regulation of Secondary Cell Wall Development by cortical Microtubules During Tracheary Element Differentiation in *Arabidopsis* Cell Suspensions." *Plant Physiology* 137(3): 1027–36.

[121] Mao, Guojie; Buschmann, Henrik; Doonan, John, H.; Lloyd, Clive, W. 2006. "The Role of MAP65-1 in Microtubule Bundling During *Zinnia* Tracheary Element Formation." *Journal of Cell Science* 119(4): 753–58.

[122] Mitsuda, Nobutaka; Seki, Motoaki; Shinozaki, Kazuo; Ohme-Takagi, Masaru. 2005. "The NAC Transcription Factors NST1 and NST2 of *Arabidopsis* Regulate Secondary Wall Thickenings and Are Required for Anther Dehiscence." *Plant Cell* 17(11): 2993–3006.

[123] Kubo, M.; Udagawa, M.; Nishikubo, N.; Horiguchi, G.; Yamaguchi, M; Ito, J; Mimura, T.; Fukuda, H.; Demura, T. 2005. "Transcription Switches for Protoxylem and Metaxylem Vessel Formation." *Genes & Development*, 19(16): 1855–60.

[124] Pesquet, Edouard; Korolev, Andrey, V.; Calder, Grant; Lloyd, Clive, W. 2010. "The Microtubule-Associated Protein AtMAP70-5 Regulates Secondary Wall Patterning in *Arabidopsis* Wood Cells." *Current Biology* 20(8): 744–49.

[125] Keech, O.; Pesquet, E.; Gutierrez, L.; Ahad, A.; Bellini, C.; Smith, S. M.; Gardeström, P. 2010. "Leaf Senescence Is Accompanied by an Early Disruption of the Microtubule Network in *Arabidopsis*." *Plant Physiology* 154(4): 1710–20.

[126] Smertenko, Andrei, P.; Bozhkov, Peter, V.; Filonova, Lada, H.; Von Arnold, Sara and Hussey, Patrick, J. 2003. "Re-Organisation of the Cytoskeleton During Developmental Programmed Cell Death in *Picea Abies* Embryos." *Plant Journal* 33(5): 813–24.

[127] Fass, E.; Shvets, E.; Degani, I.; Hirschberg, K.; Elazar, Z. 2006. "Microtubules Support Production of Starvation-Induced Autophagosomes But Not Their Targeting and Fusion With Lysosomes." *Journal of Biological Chemistry* 281(47): 36303–16.

[128] Kimura, Shunsuke; Noda, Takeshi; Yoshimori, Tamotsu. 2008. "Dynein-Dependent Movement of Autophagosomes Mediates Efficient Encounters With Lysosomes." *Cell Structure and Function* 33(1): 109–22.

[129] Yamamoto, Midori; Suzuki, Satoshi, O.; Himeno, Masaru. 2010. "The Effects of Dynein Inhibition on the Autophagic Pathway in Glioma Cells." *Neuropathology* 30(1): 1–6.

[130] Köchl, Robert; Hu, Xiao; Wen, Chan; Edmond, Y., W.; Tooze, Sharon, A. 2006. "Microtubules Facilitate Autophagosome Formation and Fusion of Autophagosomes with Endosomes." *Traffic* 7(2): 129–45.

[131] King, Stephen. M. 2002. "Dyneins Motor on in Plants." *Traffic* 3(12): 930–1.

[132] Lee, Yuh-Ru, Julie and Liu, Bo. 2004. "Cytoskeletal Motors in *Arabidopsis*. Sixty-One Kinesins and Seventeen Myosins." *Plant Physiology* 136(4): 3877–83.

[133] Reddy, Anireddy, Sn. and Day, Irene, S. 2001. "Kinesins in the *Arabidopsis* Genome: A Comparative Analysis Among Eukaryotes." *BMC Genomics* 2(1): 2. DOI:10.1186/1471-2164-2-2.

[134] Jahreiss, Luca; Menzies, Fiona, M.; Rubinsztein, David, C. 2008. "The Itinerary of Autophagosomes: From Peripheral Formation to Kiss-And-Run Fusion With Lysosomes." *Traffic* 9(4): 574–87.

[135] Ketelaar, Tijs; Voss, Christiane; Dimmock, Simon, A.; Thumm, Michael; Hussey, Patrick, J. 2004. "*Arabidopsis* Homologues of the Autophagy Protein Atg8 Are a Novel Family of Microtubule Binding Proteins." *FEBS Letters* 567(2-3): 302–6.

[136] Toyooka, K.; Moriyasu, Y.; Goto, Y.; Takeuchi, M.; Fukuda, H.; Matsuoka, K. 2006. "Protein Aggregates Are Transported to Vacuoles By a Macroautophagic Mechanism in Nutrient-Starved Plant Cells." *Autophagy* 2(2): 96–106.

[137] Thompson, Allison, R.; Doelling, Jed, H.; Suttangkakul, Anongpat; Vierstra, Richard, D. 2005 "Autophagic Nutrient Recycling in *Arabidopsis* Directed By the ATG8 and ATG12 Conjugation Pathways 1." *Plant Physiology* 138(4): 2097–110.

[138] Chung, T.; Suttangkakul, A.; Vierstra, R., D. 2009. "The ATG Autophagic Conjugation System in Maize: ATG Transcripts and Abundance of the ATG8-Lipid Adduct Are Regulated By Development and Nutrient Availability." *Plant Physiology* 149(1): 220–34.

[139] Yoshimoto, Kohki; Hanaoka, Hideki; Sato, Shusei; Kato, Tomohiko; Satoshi, Tabata; Takeshi, Noda; Yoshinori, Ohsumi. 2004. "Processing of ATG8s, Ubiquitin-Like Proteins, and Their Deconjugation By ATG4s Are Essential for Plant Autophagy." *Plant Cell* 16(11): 2967–83.

[140] Slavikova, S.; Ufaz, S.; Avin-Wittenberg, T.; Levanony, H.; Galili, G. 2008. "An autophagy-associated Atg8 Protein Is Involved in the Responses of *Arabidopsis* Seedlings to Hormonal Controls and Abiotic Stresses." *Journal of Experimental Botany* 59: 4029–43.

[141] Slávíková, Silvia; Shy, Galia; Yao, Youli; Glozman, Rina; Levanony, Hanna; Pietrokovski, Shmuel; Elazar, Zvulun; Galili, Gad. 2005. "The Autophagy-Associated Atg8 Gene Family Operates Both Under Favourable Growth Conditions and Under Starvation Stresses in *Arabidopsis* plants." *Journal of Experimental Botany* 56(421): 2839–49.

[142] Xiao, Shi and Chye, Meelen. 2010. "The *Arabidopsis thaliana* ACBP3 Regulates Leaf Senescence By Modulating Phospholipid Metabolism and ATG8 Stability." *Autophagy* 6(6): 802–4.

[143] Chung, Taijoon; Phillips, Allison, R.; Vierstra, Richard, D. 2010. "ATG8 Lipidation and ATG8-Mediated Autophagy in *Arabidopsis* Require ATG12 Expressed From the Differentially Controlled ATG12A And ATG12B loci." *Plant Journal* 62(3): 483–93.

[144] Avin-Wittenberg, Tamar; Honig, Arik; Galili, Gad. 2012. "Variations on a Theme: Plant Autophagy in Comparison to Yeast and Mammals." *Protoplasma* 249(2): 285–99.

[145] Fujioka, Y.,;Noda, N. N.; Fujii, K.; Yoshimoto, K.; Ohsumi, Y.; Inagaki, F. 2008. "In Vitro Reconstitution of Plant Atg8 and Atg12 Conjugation Systems Essential for Autophagy." *Journal of Biological Chemistry* 283(4): 1921–28.

[146] Liu, Chao; Ma, Haijie; Wu, Jiaxue; Huang, Qiang; Liu, Jun, O.; Yu, Long. 2013. "Arginine68 is an Essential Residue for the C-Terminal Cleavage of Human Atg8 Family Proteins." *BMC Cell Biology* 14(1): 27. DOI: 10.1186/1471-2121-14-27.

[147] Honig, Aric; Avin-Wittenberg, Tamar; Ufaz, Shai; Galili, Gad. 2012. "A New Type of Compartment, Defined By Lant-Specific Atg8-Interacting Proteins, is induced Upon Exposure of *Arabidopsis* Plants to Carbon Starvation." *Plant Cell* 24(1): 288–303.

[148] Michaeli, Simon,; Honig, Arik; Levanony, Hanna; Peled-Zehavi, Hadas; Galili, Gad. 2014. "*Arabidopsis* ATG8-INTERACTING PROTEIN1 Is Involved in Autophagy-Dependent Vesicular Trafficking of Plastid Proteins to the Vacuole." *Plant Cell* 26(10): 4084–101.

[149] Pankiv, Serhiy; Clausen, Terje, Høyvarde; Lamark, Trond; Brech, Andreas; Bruun, Jack-Ansgar; Outzen, Heidi; Øvervatn, Aud; Bjørkøy, Geir; Johansen, Terje. 2007. "p62/SQSTM1 Binds Directly to Atg8/LC3 to Facilitate Degradation of Ubiquitinated Protein Aggregates By Autophagy." *Journal of Biological Chemistry* 282(33): 24131–45.

[150] Kirkin, Vladimir; McEwan, David, G.; Novak, Ivana; Dikic, Ivan. 2009. "A Role For Ubiquitin in Selective Autophagy." *Molecular Cell* 34(3): 259–69.

[151] Zientara-Rytter, Katarzyna and Sirko, Agnieszka. 2014. "Significant Role of PB1 and UBA Domains in Multimerization of Joka2, a Selective Autophagy Cargo Receptor From Tobacco." *Frontiers in Plant Science* 5: 1–13.

[152] Janke, Carsten and Chloë, Bulinski, Jeannette. 2011. "Post-Translational Regulation of the Microtubule Cytoskeleton: Mechanisms and Functions." *Nature Reviews, Molecular Cell Biology* 12(12): 773–86.

[153] Xie, Rui; Nguyen, Susan; McKeehan, Wallace, L.; Liu, Leyuan. 2010. "Acetylated Microtubules Are Required For Fusion of Autophagosomes With Lysosomes." *BMC Cell Biology* 11(1): 89. DOI: 10.1186/1471-2121-11-89.

[154] Aström, Helena. 1992. "Acetylated Alpha-Tubulin in the Pollen Tube Microtubules." *Cell Biology International Reports* 16(9): 871–81.

[155] Blume, Yaroslav; Smertenko, Andrey; Ostapets, Nickolay; Viklický, Vladimır; Dráber, Pavel. 1997. "Post-Translational Modifications of Plant Tubulin." *Cell Biology International* 21(12): 917–20.

[156] Smertenko, Andrey; Blume, Yaroslav; Viklický, Vladimír; Opatrný, Zdeněk; Dráber, Pavel;. 1997. "Post-Translational Modifications and Multiple Tubulin Isoforms in *Nicotiana tabacum* L. Cells." *Planta* 201(3): 349–58.

[157] Nakagawa, Umihito; Suzuki, Daisuke; Ishikawa, Mai; Sato, Hideki; Kamemura, Kazuo; Imamura, Aya. 2013. "Acetylation of α-Tubulin on Lys40 Is a Widespread Post-Translational Modification in Angiosperms." *Bioscience, Biotechnology, and Biochemistry* 77(7): 1602–5.

[158] Huang, Rong Feng and Lloyd, Clive, W. 1999. "Gibberellic Acid Stabilises Microtubules in Maize Suspension Cells to Cold and Stimulates Acetylation of α-Tubulin." *FEBS Letters* 443: 317–20.

[159] Wang, Wei; Vignani, Rita; Scali, Monica; Sensi, Elisabetta; Cresti, Mauro. 2004. "Post-Translational Modifications of Alpha-Tubulin in *Zea mays* L. Are Highly Tissue Specific." *Planta* 218: 460–5.

[160] Nakagawa, Umihito; Kamemura, Kazuo; Imamura, Aya. 2013. "Regulated Changes in the Acetylation of α-Tubulin on Lys40 During Growth and Organ Development in Fast Plants, *Brassica rapa* L." *Bioscience, Biotechnology, and Biochemistry* 77(11): 2228–33.

[161] Xiong, Xue; Xu, Deyang; Yang, Zhongnan; Huang, Hai; Cui, Xiaofeng. 2013. "A Single Amino-Acid Substitution at Lysine 40 of an *Arabidopsis thaliana* α-Tubulin Causes Extensive Cell Proliferation and Expansion Defects." *Journal of Integrative Plant Biology* 55(3): 209–20.

[162] Eleftheriou, Eleftherios, P.; Adamakis, Ioannis-Dimosthenis, S.; Michalopoulou, Vasiliki, A. 2015. "Hexavalent Chromium-Induced Differential Disruption of Cortical microtubules in Some Fabaceae Species Is Correlated With Acetylation of α-Tubulin." *Protoplasma*. http://link.springer.com/10.1007/s00709-015-0831-4.

[163] Gzyl, Jarosław; Chmielowska-Bąk, Jagna; Przymusiński, Roman; Gwóźdź, Edward A. 2015. "Cadmium Affects Microtubule Organization

and Post-Translational Modifications of Tubulin in Seedlings of Soybean (Glycine max L.)." *Frontiers in Plant Science* 6: 1–13. http://journal.frontiersin.org/article/10.3389/fpls.2015.00937.

[164] Lytvyn, Dmytro I.; Yemets, Alla I.; Blume, Yaroslav B. 2013. "Development of Autophagy in BY-2 (*Nicotiana tabacum*) Cells Is Accompanied by Acetylation of α-Tubulin." *Proceedings of the National Academy of Sciences of Ukraine* 5: 179–85 (in Ukr.).

[165] Yoshiyama, Kaoru O.; Kimura, Seisuke; Maki, Hisaji; Britt, Anne, B.; Umeda, Masaaki. 2014. "The Role of SOG1, a Plant-Specific Transcriptional Regulator, in the DNA Damage Response." *Plant Signaling & Behavior* 9: e28889. DOI: 10.4161/psb.28889.

In: Programmed Cell Death in Plants and Animals ISBN: 978-1-63484-505-2
Editor: Josephine Rice 

*Chapter 2*

# MODULATING REGULATED CELL DEATH: THE VIRUS WAY TO INFLUENCE CELL FATE, SURVIVE AND PERSIST

***Marc Desforges*[*], *Mathieu Meessen-Pinard and Pierre J. Talbot*[†]**
Laboratory of Neuroimmunovirology, INRS-Institut Armand-Frappier, Université du Québec, Laval, Québec, Canada

## ABSTRACT

In order to thwart a viral infection, the host has developed an integrated defense network that includes innate and adaptive immunity as well as regulated cell death (RCD), which appears to represent an ancestral defence mechanism for the host. Indeed, RCD is important to control several aspects of the cell fate in various situations, from embryonic development and adult tissue homoeostasis development, to stress response, including viral infection, where different but complementary highly regulated arms of the complex RCD process exist and can be activated, with various consequences. In order for a host to cope with a viral infection, RCD can play a critical role in the elimination of virus-infected cells and/or to control the cell-to-cell viral dissemination to protect the whole organism. During the pathogen-host co-evolution,

[*] Co-Corresponding Author: E-mail:marc.desforges@iaf.inrs.ca (MD).
[†] Co-Corresponding Author: E-mail: pierre.talbot@iaf.inrs.ca (PJT).

viruses have developed various strategies to counteract defenses of the host and, among other things, they have acquired the capacity to subvert host-cell RCD. For some viruses, disabling host-cell RCD, at least for some time, could represent an important step in the replication cycle as an effort to prolong their own survival and possibly establish a persistent infection, which can lead to chronic diseases. On the other hand, other viruses may take advantage of RCD to facilitate shedding and disseminate more efficiently in the host or to prevent the presentation of viral antigens to the immune cells. By encoding numerous proteins that either are homologs to cellular apoptosis-regulatory factors or interfere with cellular proteins that control RCD-related pathways, several different viruses are able to induce or modulate the different cellular pathways that control RCD. Virus modulation of cell death participates in the pathological process associated with several human diseases; therefore understanding mechanisms and consequences of virus interactions with the host cell-death machinery is essential to understand potentially pathologically relevant consequences that will help in the design of intervention strategies and the development of antiviral therapies.

## 1. INTRODUCTION

The first hypothesis relating pathologies of unknown etiology to a "new type" of infectious agent were proposed around 1840. However, the "true" nature of these agents was only demonstrated by Montreal-born Félix d'Hérelle during the First World War [1]. Since then, viruses have been found in all living organisms, where they are associated with different forms of infections and pathologies. Not being able to replicate on their own, viruses are obligatory intracellular parasites that must use a host-cell to replicate and spread. Even though most viruses possess a rudimentary structure, accumulating evidence demonstrates that even the simplest ones have evolved and developed several ways to interact with their host, including humans.

On the other hand, to cope with the process of a viral infection, the host has developed an integrated defense network that includes innate and adaptive immunity in which several types of cells cooperate to counteract infection. The initial "innate" phase starts with the direct detection of viral nucleic acids or antigens, which will lead to production of various cytokines including type 1 interferon (IFN-1) and eventual involvement of non-immune (epithelial and endothelial cells) and immune cells such as monocytes/macrophages, dendritic cells (DC), natural killers (NK and NKT), and polymorphonuclear leukocytes

(PMN), including neutrophils [2, 3, 4]. Innate immune cells, mainly through the DC, also serves to orchestrate the implementation of the downstream more specific adaptive immunity, which will engage the diverse B and T lymphocytes that will participate in eventual viral clearance and termination of infection [5]. Another type of host "more ancestral" defence mechanism does exist in the form of a regulated cell death (RCD), which is important to control various aspects of the cell fate in several different situations, from embryonic development and adult tissue homeostasis development, to stress response, including viral infection [6].

As early as 1842, Carl Vogt, while working on the development of the amphibian nervous system, suggested that cell death could be regulated. More than a century later, in the 1950's, the embryologist Gluksmann described a process of cell death associated with specific morphological modifications during development before Lockshin and Williams (1965) presented their work on insects morphogenesis and J. R. Tata (1966) demonstrated that the loss of tadpole tail was associated with an active cell death process as reviewed in [7].

Nevertheless, the official recognition of an active process of regulated cell death (RCD), originates in 1972 in the United Kingdom, as Kerr, Willie and Currie suggested that the term Apoptosis (Greek for describing the loss of petal for a flower) be used to describe the underlying mechanism which provides a counterweight to mitosis in the regulation of cell development in animals [8]. Indeed, this pioneer work, and several others later during the 70's, contributed to changing the cell death paradigm, as the scientific community started to understand that cell death could be tightly regulated. Therefore, it could represent a cell response towards stress and it would be possible to develop therapeutic approaches to modulate it as needed [9, 10]. More than 40 years later, different forms of RCD are now described based on a functional/biochemical classification instead of morphological aspect of dying cells [11, 12].

Apoptosis represents the most studied and best characterized form of RCD. It is a complex and highly regulated mechanism that usually results in the activation of a special type of cystein proteases called the caspases (cystein-dependent aspartate specific proteases). By initiating a complex cascade of reactions that usually involve these caspases and several other cellular factors, including those from the B-cell lymphoma-2 (Bcl-2) family, the cell finally dies [13, 14, 15, 16, 17].

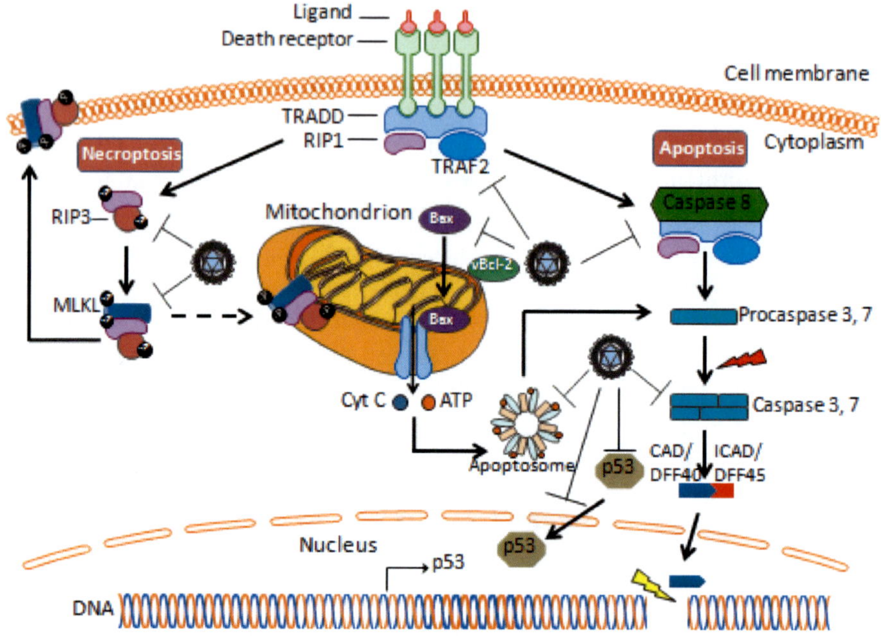


Figure 1. Pathways of regulated cell death (RCD) activated or inhibited after viral infection. Hallmarks of apoptosis, including the relocalization of the activated pro-apoptotic protein BAX (Bcl-2 associated protein X) from the cytosol to the mitochondrial membrane, cytochrome C and ATP release from mitochondria towards the cytosol, activation of caspases -3, -7, -8 and -9 and DNA fragmentation (CAD/DIFF40), are observed during a viral infection of human cells. However, viruses encode homologs of anti-apoptotic Bcl-2 (vBcl-2) and may inhibit classical caspase-dependent apoptosis by blocking Bax pro-apoptotic function at the mitochondria and/or upstream (cell death receptors) and/or downstream of caspase activation. Viruses may also inactivate the transcriptional factor p53, which is involved in active transcription of pro-apoptotic genes and inhibit apoptosis. Necroptosis may also be activated during a viral infection and induces the formation of a complex comprising phosphorylated RIP1 and RIP3, downstream of the death receptor family. This complex binds and phosphorylate (thus activating) the pseudokinase MLKL (mixed lineage kinase domain-like) to form the necrosome. This new complex is relocalized to plasma membrane (and potentially) to organelle membrane such as mitochondria, to disrupt the structure integrity, leading to the release of the intracellular contents. Viruses may also interfere in the RIP1/3 complex or activation of MLKL. Red lightning represents the cleavage of effector inactive procaspases into active caspases and yellow lightning represents the site of DNA fragmentation mediated by CAD/DFF40.

Unlike apoptosis, necrosis has historically been considered an accidental form of cell death due to physical injury or chemical insult. However, the work of Laster and colleagues, published in the late 1980's, started to challenge this paradigm [18, 19]. Indeed, almost three decades later, accumulating evidence now demonstrates that, like apoptosis, necrosis can also be tightly regulated, in certain circumstances initiated by specific signals and controlled by specific genes [20], thus making it a programmed phenomenon, which is now referred to as necroptosis [19].

The modern scientific literature, influenced in part by the Nomenclature Committee on Cell Death (NCCD), now describes several other forms of RCD, such as Parthanatos (associated with the poly(ADP-ribose) polymerase 1 (PARP1) and the apoptosis-inhibiting factor (AIF)) and pyroptosis (mainly associated with caspase-1 activation and production of IL-1b in fever) based on a functional classification [11, 12, 21]. However, as they currently represent the type of cell death mostly associated with viral infection of the host cell, the current chapter will only focus on the apoptotic and necroptotic mechanisms.

No matter the form of RCD induced in the host-cell in response to a viral infection, viruses exert a multifaceted control of the phenomenon (Figure 1). Indeed, they can prevent or at least slow down the killing of infected cells, allowing their replication cycle to proceed until completion. On the other hand, once infection has allowed an efficient replication, viruses can also trigger RCD, thereby favoring dissemination of new infectious virions.

## 2. Apoptosis

Apoptosis takes place in three stages; (1) the stimulus or initial signal, (2) the execution phase associated with morphological changes and (3) the formation of membrane-bound apoptotic bodies, which will be engulfed by surrounding cells through phagocytosis [20, 22].

The cellular apoptotic machinery can be triggered by either extracellular stress signals sensed by specific receptors integrated in the cell membrane (extrinsic apoptosis) or intracellular stress (intrinsic apoptosis), which involves the mitochondria and activates specific cellular factors that converge to trigger activation of effector caspases. Extrinsic apoptosis is activated following attachment of ligands, such as TNF$\alpha$, FasL, and TRAIL, to specific receptors (TNFR1, Fas, TRAIL-R1), which leads to a conformational change in the receptor that induces the assembly of intracellular proteins and to the activation of the initiator caspase-8 [23, 24, 25, 26]. On the other hand, intrinsic apoptosis is triggered by a variety of intracellular stress signals that activate pro-apoptotic cellular factors that impinge on the integrity of the mitochondria [14, 27]. In some cell types, the extrinsic pathway interacts with the mitochondria to amplify the apoptotic signal [27, 28, 29].

Several kinds of stress can initiate a pro-apoptotic signal where caspases play a pivotal role in the process. Their function is regulated by different proteins of the inhibitors of apoptosis (IAPs) and of the Bcl-2 families, the latter comprising both pro- and anti-apoptotic members [30, 31, 32]. Pro-apoptotic members such as Bax and Bak proteins are key regulators of the apoptotic pathway by inserting themselves in the mitochondrial outer membrane where they oligomerize to form channels (MAC; mitochondrial apoptosis-induced channel) or pores that induces permeabilization of the mitochondrial outer membrane (MOMP) and allow the release of different effector proteins, including cytochrome c and Smac/DIABLO, from the mitochondrion's intermembrane space towards cytosol [30, 33]. Effector caspases (-3 and -7) are therefore activated and can cleave numerous cytosolic and nuclear substrates, including the inactive ICAD/DFF45 and CAD/DFF40 complex, which is cleaved by caspase-3. The CAD, a double-stranded DNA-specific endonuclease, then cleaves DNA at internucleosomal regions. Other pro-apoptotic mitochondrial proteins such as Apoptosis Inducing Factor (AIF) and EndoG can also participate in DNA fragmentation in a caspase-independent manner [34, 35]. During the cell steady-state, when no apoptotic signal arises, IAP proteins interact with caspases and prevent their activation and the anti-apoptotic factors of the Bcl-2 family act as gatekeepers that prevent the oligomerization of pro-apoptotic factors and therefore allow cell survival [13, 30, 32].

## 2.1. Induction of Apoptosis Following Viral Infection

Numerous and highly diversified viruses, from small RNA viruses to large DNA viruses, induce apoptosis during infection of susceptible cells (Table 1). In the situation of a viral infection, apoptosis may have different types of consequences. One is to limit the inflammatory response [22, 36], a situation which can be beneficial to the virus for evading the immune system of the host more efficiently. In some cases, apoptosis may even enable virus spread, as it has been suggested for HIV and other viruses which were suggested to induce this mechanism after the completion of their replication cycle [6, 14, 37, 38].

Some viruses have also acquired the ability to induce RCD in uninfected cells, including immune cells, another condition that may favor viral survival and spreading [6, 27]. However, cell death itself may also contribute to and enhance virus-induced pathogenesis [39, 40]. In the end, the modulation of this complex apoptotic mechanism can sometimes be profitable to the virus and deleterious to the host. Nevertheless, apoptosis also truly represent a defense mechanism for the host, limiting viral proliferation in the infected organism [41, 42]. Hence several different viruses have developed strategies to overcome premature apoptotic cell death [43], underlining the fact that the phenomenon is part of the defense mechanism for the host [22, 42, 44]. Furthermore, if apoptosis is not induced correctly or not induced at all in a context where it should, it may allow the establishment of a persistent infection by a virus that should have been eliminated otherwise. Moreover, modulation of apoptosis has been described as a viral strategy that can also lead to the establishment of viral persistence [20, 45, 46, 47, 48].

**Table 1. Examples of virus activating apoptosis**

| Virus | Viral protein | Host | Cellular factors involved | References |
|---|---|---|---|---|
| DNA VIRUSES | | | | |
| **ADV** | E1A | Rat | Activation of p53 | [49] |
| | | Mouse | Increase sensibility to apoptosis related to TNFα | [50] |
| | E4orf4 | Human | Activation of PP2A | [36] |
| **HPV** | E7 | Mouse | Activation of p53 | [51] |
| **EBV** | EBNA3C | Rat | Association to pRb and cellular cycle progression | [52] |
| | EBNA1 | Human | Apoptosis activated when expressed | [53] |
| **HBV** | HBx | Human | Increased activity of p53 and c-myc and interaction with VDAC | [54, 55]. |
| RNA VIRUSES | | | | |
| **HIV-1** | Tat | Human | Increased of FASL expression | [56] |
| | Vpr | Human | Arrested cell cycle in G2 and interaction with ANT | [57, 58] |
| | Gp160 | Human | Increased of intracellular $Ca^{2+}$ | [59] |
| **SINV** | E1, E2 | Human | Increased activity of NF-kB in neurons | [60] |
| **HCV** | Core | Human, mouse | Association with cytoplasmic region of TNFR | [61] |

**Table 1. (Continued).**

| Virus | Viral protein | Host | Cellular factors involved | References |
|---|---|---|---|---|
| **NDV** | HN | Hamster | Induction of IFN production and TRAIL expression | [62] |
| **VSV** | M | Mouse | Unknown mechanism | [63, 64] |
| **RABV** | Glycoprotein | Mouse | Activation of caspase-8, 3, cleavage of PARP and translocation of AIF from mitochondria to nucleus | [65] |
| **WNV** | NS3 | Human, mouse | Initiation of ER stress leading to PERK, ATF6 and CHOP activation, activation of caspase-3, cleavage of PARP and growth arrest | [66, 67] |
| | Capsid protein | Human, mouse | Sequestering HDM2 (negative regulator of p53) and then activate Bax | [68] |
| **SFV** | Unknown | Mouse | Activation of Bak, Bid and caspase-8 | [69] |
| **JEV** | NS3 | Human | Activation of caspase-3 and 9 | [70, 71] |
| **TMEV** | Leader protein | Mouse | Activation of Bax and caspase-3, 9 | [72] |
| **DENV** | NS5 | Human | Interaction with Daxx that upregulates CCL5 production and Fas death receptor | [73, 74] |
| | Capsid protein | Human | Up-regulating of RIP2, caspase-8, 1 and 3 expression | [75] |
| | Unknown | Human | Activation of XAF1 that upregulates caspase-8, 9, 3 and PARP expression | [76] |
| **MERS-CoV** | Unknown | Human | Activation of caspase-3 | [77] |
| | Unknown | Human | Activation of caspase-3, 8 and 9 | [78] |
| **IAV** | Nucleoprotein | Human | Interaction with RNF43 that stabilizes and activates p53 | [79] |
| **CVB** | $3C^{Pro}$ | Human | Cleavage of Unc93b that activates apoptosis | [80] |
| **SARS-CoV** | M protein | Human | Activation of caspase-8 and 9 | [81] |

| Virus | Viral protein | Host | Cellular factors involved | References |
|---|---|---|---|---|
| **TGEV** | N protein | Swine | Accumulation of p53 and p21, Bax translocation to mitochondria, activation of caspase-3 | [82] |
| | | Swine | Downregulation of MDM2, promote phosphorylation of p53, activation of Bax | [83] |

Legends (Table 1-3): ADV, Adenovirus; AIF, Apoptosis-inducing factor; ANT, Adenine nucleotide translocator; CVB, Coxsackievirus B; Daxx, Death protein 6; DENV, Dengue virus; EBV, Epstein-Barr virus; ER, Endoplasmic reticulum; HCMV, Human cytomegalovirus; HCoV, Human coronavirus; HBV, Hepatitis B virus; HCV, Hepatitis C virus; HHV-8, Kaposi's sarcoma-associated virus; HIV-1, Human immunodeficiency virus 1; HPV, Human papillomavirus; HSV-1, 2, Herpes-simplex virus 1, 2; HTLV, Human T-lymphotropic virus; IAV, Influenza A virus; JEV, Japanese encephalitis virus; MERS-CoV, Middle East respiratory syndrome coronavirus; MCMV, Murine cytomegalovirus; MYXV, Myxomavirus; NDV, Newcastle disease virus; PARP, Poly-ADP ribose polymerase; pRb, retinoblastoma protein; PLPs, Papain-like protease; RABV, Rabies virus; SARS-CoV, Severe acute respiratory syndrome coronavirus; SFV, Semliki Forest virus; SINV, Sindbis virus; TGEV, Transmissible gastroenteritis virus; TMEV, Theiler's murine encephalomyelitis virus; TNFα, Tumor-necrosis factor alpha; VACV, Vaccinia virus; VSV, Vesicular stomatitis virus; WNV, West Nile virus.

## 2.2. Modulation of Apoptosis by Viruses

As stated above, a viral infection may induce the apoptotic machinery in the host cell and this very often serves as a defense mechanism, which will help to eliminate the invader or at least control its dissemination in order to protect the entire organism. Therefore, it is not surprising that during the pathogen-host co-evolution, viruses have developed various strategies to counteract this complex form of RCD (Table 2).

For that purpose, several of these pathogens encode proteins, which are able to modulate apoptosis. Among these viruses are human papilloma virus (HPV), hepatitis C virus (HCV), human T-cell leukemia virus (HTLV-1), Epstein Barr virus (EBV), Kaposis's sarcoma virus (KSHV or HHV-8), human adenoviruses (Ad), and simian virus 40 polyoma virus (SV40) [22, 84]. Several of these viruses have developed strategies that block specific pro-apoptotic factors. The tumor suppressor factor p53 is one target of choice for

SV40, Ad5, EBV and other herpes viruses, HPV and hepatitis B virus (HBV) and several other viruses, which all have elaborated strategies to inactivate or degrade p53 [85, 86, 87, 88, 89, 90]. The apoptotic regulators of theBcl-2 protein family represent a second cellular factor targeted by viruses to alter regulated cell-death and promote cell survival. In fact, some viruses, like Adenoviruses, EBV, Herpes virus saimiri, HCMV, and KHSV have adapted the mechanism of inhibiting cell-death in the host by encoding homologs of Bcl-2 (vBcl-2) in their own genome, while others like HCV or HTLV-1 exhibit a strategy aimed at the induction of overexpression of Bcl-2 or Bcl-xL [22, 91, 92, 93, 94, 95, 96, 97]. MicroRNAs (miRNAs) are small RNAs that play important roles in the regulation of gene expression at the post-transcriptional level in eukaryotic cells. Viral microRNAs also exist and can manipulate the host immune response and prevent apoptosis in infected cells by controlling the expression of cellular factors involved in these processes [98, 99].

**Table 2. Examples of virus interfering with apoptosis**

| **Virus** | **Viral protein** | **Host** | **Cellular factors involved** | **References** |
|---|---|---|---|---|
| DNA VIRUSES | | | | |
| **HCMV** | vICA (UL36) | Human | Inhibition of caspase-8 activity | [105] |
| | vMIA (UL37x1) | Human | Sequestering Bax at the mitochondria | [106, 107] |
| | UL38 | Human | Unknown, modulation of mTOR | [108] |
| **HHV-8** | K7 | Human | Binding to Bcl-2 and inhibition of caspase-3 activity | [109] |
| | K15 | Human | Regulation of JNK and ERK pathways and NF-kB and AP1 transcription factors | [110] |
| **VACV** | N1L | Human | Viral Bcl-2 homologue | [111] |
| | F1L | Human | Prevent interaction and activation of Bak and Bim | [112, 113] |
| | E3L, K3L | Monkey | Inhibition of PKR and RNAse L pathways | [114] |
| | B13R | Mouse | Inhibitor of caspase (serpin) | [115] |
| **Cowpox** | CrmA | Human | Inhibitor of caspase (serpin) | [116] |
| | CrmB, C | Human | TNFR analogues | [117] |
| **HSV-1** | $\gamma_1$34.5 | Human | eIF2$\alpha$ dephosphorylation | [118] |
| | US3 | Rabbit | Inhibition of Bad | [119] |

| Virus | Viral protein | Host | Cellular factors involved | References |
|---|---|---|---|---|
| **EBV** | LMP-1 | Human | Interfering with TRAF and TRADD pathways, increase Bcl-2 expression | [120, 121] |
| | EBNA4 | Human | Increased Bcl-2 expression | [122] |
| | BHRF-1 | Monkey | Bcl-2 analogue | [123] |
| **Adeno-virus** | E1B-19K | Human | Interfere with death receptor and analogue of Bcl-2 | [124] |
| | E1B-55K | Human | Inactivation of p53 | [125] |
| **Polyoma-virus** | MT | Rat | Interaction with Akt and inactivation of Bad | [126] |
| **MYXV** | M11L | Mouse | Viral Bcl-2 homologue and interfere in Bax activation | [127] |
| | Serp-2 | Rabbit | Inhibitor of caspase (serpin) | [128] |
| **MCMV** | vICA (M36) | Mouse | Interaction and inhibition of caspase-8 | [129] |
| | vMIA (M38.5) | Human | Inhibition of Bax | [130] |
| | vIBO (M41.1) | Human, mouse | Inhibition of Bak | [131] |
| RNA VIRUSES | | | | |
| **HCV** | NS2 | Human | Inhibition of Cyt C release and induction of DFF like effector | [132] |
| | E2 | Human | Inhibition of TRAIL-induced apoptosis | [133] |
| **Morbilli-virus** | V protein | Human | Sequestering p73 (analogue of p53) | [134] |
| **HIV** | Tat | Human | Increased expression of Bcl 2 | [135] |
| **HBV** | HBx | Human | Interaction with and inactivation of p53 | [136] |
| **Reovirus** | σ3 | Monkey | Inhibition of PKR pathway | [137] |
| **IAV** | NS1 | | Inhibition of PKR pathway | [138] |
| **SARS-CoV** | Enveloppe protein | Human, monkey | Reducing ER stress (IRE-1 only) | [139] |
| **HTLV** | Tax | Human | Decrease of Bim and Bid expression | [140] |
| **HCoV** | PLPs | Human | Stabilization of MDM2 leading to p53 degradation | [141] |

Another type of modulation of apoptosis by viruses may in fact be induction of apoptosis in immune cells (infected or not) that would otherwise help clear infection by killing infected cells. This situation, which help viruses like HIV [100, 101], rabies [102, 103], and measles virus [104] evade the immune system, and may also help establish persistence as the immune system becomes less able to detect and clear the viral invader.

## 3. NECROPTOSIS

In contrast to apoptosis, necroptosis is caspase-independent, involves cell rounding and swelling, and terminates with the loss of membrane integrity and leakage of cytoplasmic content, which makes it an inflammatory process [142, 143]. It has recently gained a lot of attention as a regulated necrosis independent of caspases, which is engaged when apoptosis pathway is altered or inhibited [144]. Necroptosis can be activated in several ways such as activation of Toll Like Receptor 3 (TLR3), TLR4, DNA-dependent activator of IFN regulatory factors (DAI), RNA-responsive protein kinase (PKR) and death receptors [145, 146, 147]. It is characterized by a core component composed of receptor-interacting protein kinase 1 (RIP1) and/or RIP3 interacting with each other by a RIP homotypic interaction motif (RHIM). The best characterized model of necroptosis involves attachment of TNFα to its receptor (TNFR1) on the cell surface and the formation of an intracellular complex which comprises RIP1, TRADD, TRAF2 and cIAP to promote either cell survival associated with the NF-κB pathway or caspase-8 dependent apoptosis. RIP1 and RIP3 are substrate for caspase-8, therefore in the case where caspase-8 activity is inhibited, RIP1 interacts with RIP3 and is activated by phosphorylation [148, 149, 150, 151]. The RIP1-RIP3 complex then recruits mixed-lineage kinase domain-like protein (MLKL) to form the necrosome [152, 153, 154]. The MLKL translocates to the plasma membrane to disrupt its integrity, which is associated with calcium or sodium influx [155, 156]. The MLKL may also translocate to mitochondria and regulate mitochondrial fission depending of the cell type [157, 158]. Recent data, associated with the activation of TLRs and viral infection, indicate that the RIP3 protein is the central factor involved in necroptosis and that RIP1 is not an absolute prerequisite and can be replaced by other RHIM containing factors such as the TLR adaptor molecule 1 (TICAM1 also known as TRIF) or the Z-DNA binding protein 1 (ZBP1, also known as DAI) [145, 146, 159].

### 3.1. Induction of Necroptosis Following Viral Infection

Accumulating evidence indicates that necroptosis serves as a defense mechanism against infection but also as a process used by a viral pathogen to facilitate its dissemination and spreading [160, 161]. Indeed, activation and inhibition of this regulated cell death pathway can be induced by several DNA and RNA viral agents from diverse families (Table 3).

Vaccinia virus (VV), a large double-stranded DNA virus, has been known for several years to modulate immunity and apoptosis. Using the anti-apoptotic B13R viral proteins, VV is able to inhibit caspase-associated apoptosis and the cell response appears to induce necroptosis as a trap-door to counteract infection [115, 160, 162]. Using a mouse model, it was shown that this antiviral strategy significantly limits viral replication but induces a more severe pathology [150, 163]. Like VV, MCMV, the murine counterpart of the human cytomegalovirus (HCMV), member of the herpes virus family, also encodes a viral inhibitor of caspase-8-dependent apoptosis [105, 164] and can induce necroptosis as an antiviral mechanism [165]. Reovirus, a small non-enveloped double-stranded RNA virus, induces apoptotic regulated cell-death in several different cell types and involves death receptor signaling, especially the TNF-related apoptosis-inducing ligand (TRAIL) [166, 167]. Recent data indicate that Reovirus can also induce RIP-1 associated necroptosis in the murine fibroblasts cell line L-929 [168]. Using its non-structural protein 1 (nsp1), Influenza A virus (IAV), a single-stranded multi-segment RNA virus causing the flu is able to inhibit apoptosis early during infection and to induce it at later stages [169]. The IAV is also able to induce lethal death-receptor-associated necroptosis in cIAP2-deficient infected mice [170]. HIV-1, the etiologic retrovirus agent of AIDS, also induces necroptosis in infected CD4 T cells [171]. The human coronavirus strain OC43 (HCoV-OC43), a single-stranded RNA virus affecting mostly the respiratory tract, represents the most recent viral candidate for which induction of necroptosis in neuronal cell cultures appears to be an antiviral cell response. In this case, the S viral glycoprotein is implicated in induction of cell-death and genetic depletion of RIP-1 increased cell survival after infection [172]. Recent data indicate that, like apoptosis, necroptosis may represent a cellular response that limits viral infection and spread. In fact, it appears to emerge as an alternative defense mechanism against viral infection (when apoptosis is blocked by viral inhibitors) as a growing number of viruses appear to be able to manipulate it in order to help them disseminate and spread [20, 160].

**Table 3. Examples of virus interacting with necroptosis**

| Virus | Viral protein | Host | Cellular factors involved | References |
|---|---|---|---|---|
| Activation of Necroptosis | | | | |
| **VACV** | Unknown | Human | Interaction of RIP1 with RIP3, RIP1 phosphorylation | [150, 163] |
| **MCMV** | Unknown | Mouse | Increase of RIP1 mRNA | [176] |
| **Reovirus** | Unknown | Mouse | RIP1 activation, reduction of ATP level and increase in HMGB1 release | [168] |
| **HIV-1** | Tat, Enveloppe | Human | Activation of RIP1 and RIP3 | [171] |
| **IAV** | Unknown | Human | Activation of RIP1 and RIP3 | [170] |
| **TMEV** | Unknown | Human | Colocalization of RIP1 and RIP3 at the mitochondria, phosphorylation of RIP1 | [177] |
| **HSV-1** | ICP6, ICP10 | Mouse | Interaction of RIP1 with RIP3, phosphorylation of RIP1, RIP3 and MLKL | [178, 179] |
| **HCoV (OC43)** | Unknown | Human, mouse | Unknown | Meessen-Pinard M et al., 2016, J. Virol. (in submission) |
| Inhibition of Necroptosis | | | | |
| **MCMV** | vIRA (M45) | Mouse | Target RIP3 to disrupt interaction between RIP1 and RIP3 | [165, 180] |
| **HCMV** | IE1 | Human | Inhibition of RIP3 and MLKL phosphorylation | [174] |
| **HSV-1, 2** | ICP6, ICP10 | Human | Disruption of interaction between RIP1 and RIP3 | [173] |
| **CVB** | $3C^{Pro}$ | Human | Cleavage of RIP3 | [181] |

## 3.2. Modulation of Necroptosis by Viruses

Even though apoptosis has long been described as the main form of RCD related to host defense, more and more evidence indicates that necroptosis and related forms of programmed necrosis are also important in that matter as the underlying mechanisms of viral inhibition are now beginning to be unraveled. Until now, only a few large DNA viruses of the herpesvirus family have been shown to modulate necroptosis. Indeed, murine (MCMV) and human (HCMV) Cytomegalovirus, as well as Herpes Simplex virus (HSV-1 and HSV-2) have all been shown to encode viral proteins that block necroptosis in order to regulate infection for their own benefit in susceptible host cells [165, 173, 174]. Moreover, as necroptosis terminates with the loss of membrane integrity and leakage of cytoplasmic content, which makes it an inflammatory process, its inhibition may certainly help viruses evade immune response [175]. As for modulation of apoptosis, the alteration of the necroptotic process may help a virus to circumvent cell mortality as an effort to prolong their own survival and possibly establish a persistent infection.

## 4. Persistent Viral Infection

Viral infections can be of two types. In the acute form, a virus usually replicates and disseminates rapidly in order to counteract the different host defense mechanisms, which are there to control and eliminate the pathogen. In general, acute infections are rapid and usually self-limiting. Therefore, one can see the pathological consequences associated with an acute viral infection as some sort of side effect, which is deleterious to the infected individuals but also to the virus. Indeed, highly pathogenic infections are often self-limiting because of the risk of eliminating suitable host for viral transmission and perpetuation [90]. In some situations, viruses will only be partially controlled by the host and they will be able to persist in the host. These persistent infections can be either chronic, as new viral particles will continuously be produced (hepatitis B virus in humans) or latent, where the viral genome persists in different states and where viral production is occasional (herpes simplex virus HSV or human immunodeficiency virus HIV-1). Being obligate parasites, which absolutely depend on a living cell for reproduction and survival, throughout the process of pathogen-host co-evolution, viruses have therefore developed strategies to manipulate or at least bypass host defenses [6, 14].

In order to establish persistence, several viral strategies do exist. The simplest (and often more cytopathic) viruses do not necessarily possess any precise strategy encrypted in their own genome. In fact, several of those viruses are RNA viruses and they must rely on the natural selection pressure and on their extraordinary capacity of adaptation, in order to acquire favorable mutations that will make them less cytopathic, allowing the establishment of persistence in a host [182]. Viruses can also modulate the expression of cellular factors involved in the cell death machinery or in the production of type 1 interferon (IFN-1) or other cytokines in an effort to manipulate the different arms of the immune system [182, 183, 184, 185, 186]. On the other hand, and as stated above other viruses have developed sophisticated strategies to establish a persistent infection, which involves the modulation of regulated cell death. As stated above, during a viral infection, RCD can play a critical role in the elimination of virus-infected cells and/or to control the cell-to-cell viral dissemination in order to protect the organism. Therefore, controlling cellular RCD represents one important viral adaptation to limit cell stress response, which allows viral survival without compromising cell viability. Some viruses have developed various strategies to disable host-cell RCD, at least for some time as an effort to prolong their own survival and possibly establish a persistent infection. This type of infection may, in certain circumstances, lead to complex consequences often difficult to clearly identify and understand in the host. Subversion of regulated cell death, either apoptotic or necroptotic, is clearly part of the process and may, in certain circumstances, be associated with different types of animal and human pathologies.

## 5. Pathologies Associated with Persistent Viral Infections

### 5.1. Animal Persistent Infection and Related Pathologies

Several animal viruses can establish a persistent infection in their host and this situation may be associated with several different types of pathologies. Table 4, adapted from Ahmed and collaborators [187], presents different examples of such viruses.

Even though a persistent viral infection may be established in different organs within the animal, the central nervous system (CNS) represents a site of choice where immune cells have restricted access and neurodegenerative diseases may be linked to persistent viral infections in the CNS in animals. Indeed, viral induction of long term neuropathology is well described for Theiler's murine encephalitis virus (TMEV)-induced demyelination disease in mice [41, 188] and subtle neurological deficit associated with lymphocytic choriomeningitis virus (LCMV) [189]. The murine coronavirus mouse hepatitis virus (MHV) is a subspecies of the species murine coronavirus (MuCoV) [190], which represents a collection of viral strains with different tropism, including respiratory for the MHV-1 strain, and neurotropic for the MHV-JHM and MHV-A59 strains. In susceptible mice, these two latter strains of MHV induce a demyelinating disease that resembles human multiple sclerosis (MS) [191, 192, 193, 194, 195, 196] and can persist in neural cells [197, 198].

**Table 4. Examples of persistent viral infections in animal**

| Virus | Natural host | Associated disease |
|---|---|---|
| DNA VIRUSES | | |
| **Feline parvovirus** | Cat | Feline panleukopenia, cerebellar ataxia in newborn |
| **HBV** | Monkey | Encephalitis |
| **MCMV** | Mouse | Cytomegalovirus inclusion disease of multiple organs |
| **Polyomavirus** | Mouse | Tumors of multiple tissues |
| **SV40** | Monkey | Progressive multifocal leukoencephalopathy in immunosuppressed primates only |
| **DHV** | Peking duck | Hepatitis, hepatocellular carcinoma |
| **WHV** | Woodchuck | Hepatitis, hepatocellular carcinoma |
| RNA VIRUSES | | |
| **TMEV** | Mouse | CNS demyelination |
| **MHV** | Mouse | CNS demyelination |
| **Canine distemper virus** | Dog | CNS demyelination |
| **LCMV** | Mouse | Chronic glomerulonephritis |
| **Visna/maedi virus** | Sheep | Interstitial pneumonitis, CNS demyelination |
| **Caprine arthritis encephalitis virus** | Goat | Chronic arthritis, encephalomyelitis |
| **Equine infectious anemia virus** | Horse | Recurrent haemolytic anemia |
| **Avian leucosis virus** | Chicken | Leukemia |
| **Avian sarcoma virus** | Chicken | Sarcoma |
| **Murine leukemia virus** | Mouse | Leukemia |

## 5.2. Human Persistent Infection and Related Pathologies

Human pathologies are sometimes also partially related to persistent viral infections. Table 5, also adapted from Ahmed and colleagues [187] presents a list of well documented association between persisting viruses and development of human pathologies.

**Table 5. Examples of persistent viral infections in human**

| Virus | Location of persistence | Associated disease |
|---|---|---|
| DNA VIRUSES | | |
| **Adenovirus** | Adenoids, tonsils, lymphocytes | Unknown |
| **HCMV** | Kidney, salivary glands, lymphocytes? macrophages? stromal cells? | Pneumonia, retinitis |
| **EBV** | Pharyngeal epithelial cells, B cells | Infectious mononucleosis, Burkitt's lymphoma, nasopharyngeal carcinoma, non-Hodgkin's lymphoma, oral hairy leucoplakia |
| **HSV-1, 2** | Sensory ganglia neurons | Cold sores, genital herpes, encephalitis, keratitis |
| **HHV-6** | Lymphocytes | Exanthem subitum |
| **Varicella zoster virus** | Sensory ganglia neurons and/or satellite cells | Varicella, zoster |
| **HBV** | Hepatocytes, lymphocytes? macrophage? | Hepatitis, hepatocellular carcinoma |
| **HDV** | Hepatocytes | Exacerbation of chronic HBV infection |
| **Papillomavirus** | Epithelial skin cells | Papilloma, carcinoma |
| **Parvovirus B19** | Erythroid progenitor cell in bone marrow | Aplastic crisis in haemolytic anemia, chronic bone marrow deficiency |
| **Polyomavirus BK** | Kidney | Hemorrhagic cystitis |
| **Polyomavirus JC** | Kidney, oligodendrocytes in the CNS | Progressive multifocal leukoencephalopathy |
| RNA VIRUSES | | |
| **HCV** | Hepatocytes, lymphocytes? macrophages? | Hepatitis, hepatocellular carcinoma |
| **Measles virus** | Neurons and supporting cells in the CNS | Subacute sclerosis panencephalitis, measles inclusion body encephalitis |
| **Rubella virus** | CNS | Progressive rubella panencephalitis, insulin-dependent diabetes? Juvenile arthritis? |
| **HIV** | CD4 T cells, monocytes, macrophages, microglia | AIDS |
| **HTLV-1** | T cells | T-cell leukemia, tropical spastic paraparesis, polymyositis |

Legends (Table 4-5): DHV, Duck Hepatitis virus; EBV, Epstein-Barr Virus; HBV, Hepatitis B virus; HCV, Hepatitis C virus; HDV, Hepatitis D virus; HSV-1, 2, Herpes simplex virus 1, 2; HCMV, Human Cytomegalovirus; HHV-6, Human herpesvirus 6; HIV, Human immunodeficiency virus; HTLV-1, Human T-lymphotropic virus 1; LCMV, Lymphocytic choriomeningitis virus; MCMV, Murine Cytomegalovirus; MHV, Murine Hepatitis virus; TMEV, Theiler's murine encephalomyelitis; SV40, Simian virus 40; WHV, Woodchuck Hepatitis virus.

Some viruses that are able to establish a persistent infection possess oncogenic properties and it is thought that these viruses are associated with the development of 15-20% of all different forms of cancer [84]. By reprogramming host cellular pathways, including the regulated cell death machinery, viruses such as EBV, HCMV, KSHV, HTLV-1, HPV and HCV may predispose and participate to cancer development [96, 199]. Indeed, even though viruses have without a doubt been identified as etiologic agents of different tumors, infection alone is not sufficient to induce cancer, as most infected individuals will not necessarily develop tumors. For those unfortunate who will, several years can separate the initial infection and the appearance of a tumor, thus suggesting that a persisting virus, which blocks normal control of RCD, and several other factors, such as host genetic background and environment, are involved in the process [84, 96]. Depending on the type of viral persistence, inflammation, like the one observed in HCV persistent infection, is likely to be an important driver of cancer as activated inflammatory cells may promote a pro-carcinogenic microenvironment for cells which already harbor mutations that predispose an individual to cancer and auto-immunity [200, 201, 202, 203]. Persisting viruses, such as HCMV, which persist in haematopoietic progenitors in the bone marrow (and maybe in endothelial cells and neuronal progenitors) may also be opportunistic pathogens that will reactivate in immunosuppressed individuals (fetuses elderly, AIDS patients, transplant patients, intensive care patients) causing diverse health problems, important morbidity and even death [204, 205, 206].

A persistent viral infection is one of several factors that, taken together, can lead to a chronic neuropathology [207]. In several of these pathologies, a persistent infection, whereby a virus has elaborated a strategy of non-lytic replication in the infected cells, often seems to be a key factor in the triggering of illness [208]. Given the observation that several viruses induce or modulate apoptosis, it appears that this process is a key factor in the complex regulation that takes place to control the infection, including modulation from an acute infection to viral persistence. The central nervous system represents a privileged site for viral persistence and this could imply relevance to the development of neurological disorders.

Over the years, the CNS has been shown to represent a frequent site of viral infection. In fact, even though the incidence of viral infections of the CNS is not well defined, they are not uncommon occurrences in clinical practice and an increasing number of positive viral identifications are now made with the help of modern molecular diagnostics [209]. Using different routes of entry, several different viruses have been shown to be able to penetrate the CNS (neuroinvasion), where they can infect neurons and glial cells (neurotropism) and possibly induce or participate in the induction of neurological diseases (neurovirulence) [210]. In humans, a long list of viruses, for which the primary site of infection in human is not the CNS, possess these "neuroproperties" and neuroviral infection often leads to acute encephalitis, which can be fatal depending on virus tropism [211]. For example, rabies virus [212], herpes simplex virus (HSV) [213], arthropod-borne flaviviruses such as West Nile virus (WNV) or Japanese encephalitis virus (JEV) [214, 215, 216] and enteroviruses such as poliovirus (PV) and coxsackievirus (CV) [217, 218], affect millions of individuals worldwide each year and can induce encephalitis, meningitis and paralysis in humans. Respiratory viral agents also have the capacity to invade the human CNS where they will infect resident cells and potentially be neurovirulent in inducing a neuropathology. Among these recognized respiratory pathogens that can gain access to the CNS, where they can eventually cause health problems in humans including encephalitis, are the human coronaviruses [219, 220, 221].

Chronic human neurological diseases and/or sequelae may also be linked to viral infection. In acquired immunodeficiency syndrome (AIDS) dementia and related disorders, human immunodeficiency virus (HIV) induces neurodegeneration [222], which can result in motor dysfunctions and possibly cognitive impairments [223]. Progressive multifocal leukoencephalopathy (PML) is a human demyelinating disease [224, 225] where prolonged immunosuppression leads to reactivation of latent polyoma JC virus (JCV) [226]. Human T-lymphotropic virus (HTLV-1) causes progressive tropical spastic paraperesis / HTLV-1-associated myelopathy (PTSP/HAM) in 1-2% of infected individuals [227, 228] and HSV-1 and human herpes virus 6 (HHV-6) were proposed to cause or exacerbate Alzheimer's disease (AD) [229].

Coronaviruses that infect humans are not well characterized concerning their capacity to invade and infect the CNS. However, the detection of coronaviral RNA in human brain samples clearly demonstrates that these respiratory pathogens are naturally neuroinvasive in humans and suggests that they establish a persistent infection in human CNS [230]. Furthermore, these viruses are able to establish a persistent infection in human cells representative of the CNS [231, 232] and HCoV-OC43 RNA is detected for at least a year in the CNS of infected mice and a significant portion of these surviving mice exhibit abnormal reflexes shown by limb clasping, present clinical signs of decreased activity in an open field test, and have a smaller hippocampus associated with a loss of hippocampal neurons, particularly in the CA1 and CA3 layers [233]. Paralysis, neurodegeneration and demyelination are also observed in spinal cord of infected mice [234, 235]. Therefore, an apparently innocuous human respiratory pathogen may persist in the human CNS and it would therefore be possible that such a persistent infection may become a factor or co-factor of neuropathogenesis associated with long-term neurological sequelae, such as multiple sclerosis in genetically or otherwise predisposed individuals [219].

## Conclusion

Multicellular organisms respond to viral infection by activating a plethora of protective mechanisms, which represent an integrated strategic defense network that includes innate and adaptive immunity as well as regulated cell death (RCD) and conversely, viruses have developed a wide array of strategies that aim to limit the host response by manipulating underlying mechanisms to their advantage in order to influence cell fate and survive, in some cases by initiating a persistent infection which could either be silent or occasionally activated by some cellular or viral factors. Viruses and humans have co-evolved for several thousands of years and it is essential to precisely understand relevant virus-host interactions. More specifically, virus modulation of cell death participates in the pathological process associated with several human diseases; therefore studies of RCD are warranted to decipher its underlying mechanisms and their control, with the ultimate aim to clinically modulate it in patients in a beneficial way that will help in the design of intervention strategies and the development of antiviral therapies.

## References

[1] Levine J (2001). The origin of virology. In: Knipe D, Howley P, editors. Fields Virology fourth edition. Philadelphia: *Lippincott Williams and Wilkins*. pp. 3-18.

[2] Biron C, Sen G (2007) Innate responses to viral infections. In: Knipe D, Howley P, editors. Fields Virology fifth edition. Philadelphia: *Lippincott Williams and Wilkins*. pp. 249-278.

[3] Collins SE, Mossman KL (2014) Danger, diversity and priming in innate antiviral immunity. *Cytokine Growth Factor Rev* 25: 525-531.

[4] Hare D, Mossman KL (2013) Novel paradigms of innate immune sensing of viral infections. *Cytokine* 63: 219-224.

[5] Braciale T, Hahn Y, Burton D (2007) The adaptive immune response to viruses. In: Knipe D, Howley P, editors. Fields Virology fifth edition. Philadelphia: *Lippincott Williams Wilkins*. pp. 279-325.

[6] Galluzzi L, Kepp O, Morselli E, Vitale I, Senovilla L, et al. (2010) Viral strategies for the evasion of immunogenic cell death. *J Intern Med* 267: 526-542.

[7] Vaux DL, Korsmeyer SJ (1999) Cell death in development. *Cell* 96: 245-254.

[8] Kerr JF, Wyllie AH, Currie AR (1972) Apoptosis: a basic biological phenomenon with wide-ranging implications in tissue kinetics. *Br J Cancer* 26: 239-257.

[9] Wyllie AH, Kerr JF, Currie AR (1980) Cell death: the significance of apoptosis. *Int Rev Cytol* 68: 251-306.

[10] Danial NN, Korsmeyer SJ (2004) Cell death: critical control points. *Cell* 116: 205-219.

[11] Galluzzi L, Vitale I, Abrams JM, Alnemri ES, Baehrecke EH, et al. (2012) Molecular definitions of cell death subroutines: recommendations of the Nomenclature Committee on Cell Death 2012. *Cell Death Differ* 19: 107-120.

[12] Galluzzi L, Bravo-San Pedro JM, Vitale I, Aaronson SA, Abrams JM, et al. (2015) Essential versus accessory aspects of cell death: recommendations of the NCCD 2015. *Cell Death Differ* 22: 58-73.

[13] Kvansakul M, Hinds MG (2013) Structural biology of the Bcl-2 family and its mimicry by viral proteins. *Cell Death Dis* 4: e909.

[14] Galluzzi L, Brenner C, Morselli E, Touat Z, Kroemer G (2008) Viral control of mitochondrial apoptosis. *PLoS Pathog* 4: e1000018.

[15] Nunez G, Benedict MA, Hu Y, Inohara N (1998) Caspases: the proteases of the apoptotic pathway. *Oncogene* 17: 3237-3245.

[16] Earnshaw WC, Martins LM, Kaufmann SH (1999) Mammalian caspases: structure, activation, substrates, and functions during apoptosis. *Annu Rev Biochem* 68: 383-424.

[17] Hengartner MO (2000) The biochemistry of apoptosis. *Nature* 407: 770-776.

[18] Laster SM, Wood JG, Gooding LR (1988) Tumor necrosis factor can induce both apoptic and necrotic forms of cell lysis. *J Immunol* 141: 2629-2634.

[19] Vandenabeele P, Galluzzi L, Vanden Berghe T, Kroemer G (2010) Molecular mechanisms of necroptosis: an ordered cellular explosion. *Nat Rev Mol Cell Biol* 11: 700-714.

[20] Upton JW, Chan FK (2014) Staying alive: cell death in antiviral immunity. *Mol Cell* 54: 273-280.

[21] Vanden Berghe T, Linkermann A, Jouan-Lanhouet S, Walczak H, Vandenabeele P (2014) Regulated necrosis: the expanding network of non-apoptotic cell death pathways. *Nat Rev Mol Cell Biol* 15: 135-147.

[22] Teodoro JG, Branton PE (1997) Regulation of apoptosis by viral gene products. *J Virol* 71: 1739-1746.

[23] Wajant H, Pfizenmaier K, Scheurich P (2003) Tumor necrosis factor signaling. *Cell Death Differ* 10: 45-65.

[24] Seol DW, Li J, Seol MH, Park SY, Talanian RV, et al. (2001) Signaling events triggered by tumor necrosis factor-related apoptosis-inducing ligand (TRAIL): caspase-8 is required for TRAIL-induced apoptosis. *Cancer Res* 61: 1138-1143.

[25] Lettau M, Paulsen M, Schmidt H, Janssen O (2011) Insights into the molecular regulation of FasL (CD178) biology. *Eur J Cell Biol* 90: 456-466.

[26] Li J, Yin Q, Wu H (2013) Structural basis of signal transduction in the TNF receptor superfamily. *Adv Immunol* 119: 135-153.

[27] Neumann S, El Maadidi S, Faletti L, Haun F, Labib S, et al. (2015) How do viruses control mitochondria-mediated apoptosis? *Virus Res* 209: 45-55.

[28] Li H, Zhu H, Xu CJ, Yuan J (1998) Cleavage of BID by caspase 8 mediates the mitochondrial damage in the Fas pathway of apoptosis. *Cell* 94: 491-501.

[29] Luo X, Budihardjo I, Zou H, Slaughter C, Wang X (1998) Bid, a Bcl2 interacting protein, mediates cytochrome c release from mitochondria in response to activation of cell surface death receptors. *Cell* 94: 481-490.

[30] Kuwana T, Newmeyer DD (2003) Bcl-2-family proteins and the role of mitochondria in apoptosis. *Curr Opin Cell Biol* 15: 691-699.

[31] Jourdain A, Martinou JC (2009) Mitochondrial outer-membrane permeabilization and remodelling in apoptosis. *Int J Biochem Cell Biol* 41: 1884-1889.

[32] Kocab AJ, Duckett CS (2015) Inhibitor of apoptosis proteins as intracellular signaling intermediates. *FEBS J.*

[33] Ghibelli L, Diederich M (2010) Multistep and multitask Bax activation. *Mitochondrion* 10: 604-613.

[34] Kitazumi I, Tsukahara M (2011) Regulation of DNA fragmentation: the role of caspases and phosphorylation. *FEBS J* 278: 427-441.

[35] Polster BM (2013) AIF, reactive oxygen species, and neurodegeneration: a "complex" problem. *Neurochem Int* 62: 695-702.

[36] Marcellus RC, Lavoie JN, Boivin D, Shore GC, Ketner G, et al. (1998) The early region 4 orf4 protein of human adenovirus type 5 induces p53-independent cell death by apoptosis. *J Virol* 72: 7144-7153.

[37] Strack PR, Frey MW, Rizzo CJ, Cordova B, George HJ, et al. (1996) Apoptosis mediated by HIV protease is preceded by cleavage of Bcl-2. *Proc Natl Acad Sci U S A* 93: 9571-9576.

[38] Kaminskyy V, Zhivotovsky B (2010) To kill or be killed: how viruses interact with the cell death machinery. *J Intern Med* 267: 473-482.

[39] Jelachich ML, Lipton HL (1996) Theiler's murine encephalomyelitis virus kills restrictive but not permissive cells by apoptosis. *J Virol* 70: 6856-6861.

[40] Ubol S, Sukwattanapan C, Utaisincharoen P (1998) Rabies virus replication induces Bax-related, caspase dependent apoptosis in mouse neuroblastoma cells. *Virus Res* 56: 207-215.

[41] Tsunoda I, Kurtz CI, Fujinami RS (1997) Apoptosis in acute and chronic central nervous system disease induced by Theiler's murine encephalomyelitis virus. *Virology* 228: 388-393.

[42] Allsopp TE, Scallan MF, Williams A, Fazakerley JK (1998) Virus infection induces neuronal apoptosis: A comparison with trophic factor withdrawal. *Cell Death Differ* 5: 50-59.

[43] Benedict CA, Norris PS, Ware CF (2002) To kill or be killed: viral evasion of apoptosis. *Nat Immunol* 3: 1013-1018.

[44] Barber GN (2001) Host defense, viruses and apoptosis. *Cell Death Differ* 8: 113-126.

[45] Shen Y, Shenk TE (1995) Viruses and apoptosis. *Curr Opin Genet Dev* 5: 105-111.

[46] Hardwick JM (1998) Viral interference with apoptosis. *Semin Cell Dev Biol* 9: 339-349.

[47] Kalvakolanu DV (1999) Virus interception of cytokine-regulated pathways. *Trends Microbiol* 7: 166-171.

[48] Griffin DE, Hardwick JM (1999) Perspective: virus infections and the death of neurons. *Trends Microbiol* 7: 155-160.

[49] Debbas M, White E (1993) Wild-type p53 mediates apoptosis by E1A, which is inhibited by E1B. *Genes Dev* 7: 546-554.

[50] Duerksen-Hughes P, Wold WS, Gooding LR (1989) Adenovirus E1A renders infected cells sensitive to cytolysis by tumor necrosis factor. *J Immunol* 143: 4193-4200.

[51] Howes KA, Ransom N, Papermaster DS, Lasudry JG, Albert DM, et al. (1994) Apoptosis or retinoblastoma: alternative fates of photoreceptors expressing the HPV-16 E7 gene in the presence or absence of p53. *Genes Dev* 8: 1300-1310.

[52] Parker GA, Crook T, Bain M, Sara EA, Farrell PJ, et al. (1996) Epstein-Barr virus nuclear antigen (EBNA)3C is an immortalizing oncoprotein with similar properties to adenovirus E1A and papillomavirus E7. *Oncogene* 13: 2541-2549.

[53] Gregory CD, Dive C, Henderson S, Smith CA, Williams GT, et al. (1991) Activation of Epstein-Barr virus latent genes protects human B cells from death by apoptosis. *Nature* 349: 612-614.

[54] Chirillo P, Pagano S, Natoli G, Puri PL, Burgio VL, et al. (1997) The hepatitis B virus X gene induces p53-mediated programmed cell death. *Proc Natl Acad Sci U S A* 94: 8162-8167.

[55] Rahmani Z, Huh KW, Lasher R, Siddiqui A (2000) Hepatitis B virus X protein colocalizes to mitochondria with a human voltage-dependent anion channel, HVDAC3, and alters its transmembrane potential. *J Virol* 74: 2840-2846.

[56] Westendorp MO, Shatrov VA, Schulze-Osthoff K, Frank R, Kraft M, et al. (1995) HIV-1 Tat potentiates TNF-induced NF-kappa B activation and cytotoxicity by altering the cellular redox state. *EMBO J* 14: 546-554.

[57] Stewart SA, Poon B, Jowett JB, Chen IS (1997) Human immunodeficiency virus type 1 Vpr induces apoptosis following cell cycle arrest. *J Virol* 71: 5579-5592.

[58] Jacotot E, Ravagnan L, Loeffler M, Ferri KF, Vieira HL, et al. (2000) The HIV-1 viral protein R induces apoptosis via a direct effect on the mitochondrial permeability transition pore. *J Exp Med* 191: 33-46.

[59] Sasaki M, Uchiyama J, Ishikawa H, Matsushita S, Kimura G, et al. (1996) Induction of apoptosis by calmodulin-dependent intracellular Ca2+ elevation in CD4+ cells expressing gp 160 of HIV. *Virology* 224: 18-24.

[60] Ubol S, Tucker PC, Griffin DE, Hardwick JM (1994) Neurovirulent strains of Alphavirus induce apoptosis in bcl-2-expressing cells: role of a single amino acid change in the E2 glycoprotein. *Proc Natl Acad Sci U S A* 91: 5202-5206.

[61] Zhu N, Khoshnan A, Schneider R, Matsumoto M, Dennert G, et al. (1998) Hepatitis C virus core protein binds to the cytoplasmic domain of tumor necrosis factor (TNF) receptor 1 and enhances TNF-induced apoptosis. *J Virol* 72: 3691-3697.
[62] Zeng J, Fournier P, Schirrmacher V (2002) Induction of interferon-alpha and tumor necrosis factor-related apoptosis-inducing ligand in human blood mononuclear cells by hemagglutinin-neuraminidase but not F protein of Newcastle disease virus. *Virology* 297: 19-30.
[63] Kopecky SA, Willingham MC, Lyles DS (2001) Matrix protein and another viral component contribute to induction of apoptosis in cells infected with vesicular stomatitis virus. *J Virol* 75: 12169-12181.
[64] Desforges M, Despars G, Berard S, Gosselin M, McKenzie MO, et al. (2002) Matrix protein mutations contribute to inefficient induction of apoptosis leading to persistent infection of human neural cells by vesicular stomatitis virus. *Virology* 295: 63-73.
[65] Sarmento L, Tseggai T, Dhingra V, Fu ZF (2006) Rabies virus-induced apoptosis involves caspase-dependent and caspase-independent pathways. *Virus Res* 121: 144-151.
[66] Medigeshi GR, Lancaster AM, Hirsch AJ, Briese T, Lipkin WI, et al. (2007) West Nile virus infection activates the unfolded protein response, leading to CHOP induction and apoptosis. *J Virol* 81: 10849-10860.
[67] Ramanathan MP, Chambers JA, Pankhong P, Chattergoon M, Attatippaholkun W, et al. (2006) Host cell killing by the West Nile Virus NS2B-NS3 proteolytic complex: NS3 alone is sufficient to recruit caspase-8-based apoptotic pathway. *Virology* 345: 56-72.
[68] Yang MR, Lee SR, Oh W, Lee EW, Yeh JY, et al. (2008) West Nile virus capsid protein induces p53-mediated apoptosis via the sequestration of HDM2 to the nucleolus. *Cell Microbiol* 10: 165-176.
[69] Urban C, Rheme C, Maerz S, Berg B, Pick R, et al. (2008) Apoptosis induced by Semliki Forest virus is RNA replication dependent and mediated via Bak. *Cell Death Differ* 15: 1396-1407.
[70] Yang TC, Shiu SL, Chuang PH, Lin YJ, Wan L, et al. (2009) Japanese encephalitis virus NS2B-NS3 protease induces caspase 3 activation and mitochondria-mediated apoptosis in human medulloblastoma cells. *Virus Res* 143: 77-85.
[71] Yiang GT, Chen YH, Chou PL, Chang WJ, Wei CW, et al. (2013) The NS3 protease and helicase domains of Japanese encephalitis virus trigger cell death via caspasedependent and independent pathways. *Mol Med Rep* 7: 826-830.

[72] Fan J, Son KN, Arslan SY, Liang Z, Lipton HL (2009) Theiler's murine encephalomyelitis virus leader protein is the only nonstructural protein tested that induces apoptosis when transfected into mammalian cells. *J Virol* 83: 6546-6553.

[73] Khunchai S, Junking M, Suttitheptumrong A, Yasamut U, Sawasdee N, et al. (2012) Interaction of dengue virus nonstructural protein 5 with Daxx modulates RANTES production. *Biochem Biophys Res Commun* 423: 398-403.

[74] Nagila A, Netsawang J, Srisawat C, Noisakran S, Morchang A, et al. (2011) Role of CD137 signaling in dengue virus-mediated apoptosis. *Biochem Biophys Res Commun* 410: 428-433.

[75] Morchang A, Yasamut U, Netsawang J, Noisakran S, Wongwiwat W, et al. (2011) Cell death gene expression profile: role of RIPK2 in dengue virus-mediated apoptosis. *Virus Res* 156: 25-34.

[76] Long X, Li Y, Qi Y, Xu J, Wang Z, et al. (2013) XAF1 contributes to dengue virus-induced apoptosis in vascular endothelial cells. *FASEB J* 27: 1062-1073.

[77] Tao X, Hill TE, Morimoto C, Peters CJ, Ksiazek TG, et al. (2013) Bilateral entry and release of Middle East respiratory syndrome coronavirus induces profound apoptosis of human bronchial epithelial cells. *J Virol* 87: 9953-9958.

[78] Chu H, Zhou J, Wong BH, Li C, Chan JF, et al. (2015) Middle East Respiratory Syndrome Coronavirus Efficiently Infects Human Primary T Lymphocytes and Activates the Extrinsic and Intrinsic Apoptosis Pathways. *J Infect Dis*.

[79] Nailwal H, Sharma S, Mayank AK, Lal SK (2015) The nucleoprotein of influenza A virus induces p53 signaling and apoptosis via attenuation of host ubiquitin ligase RNF43. *Cell Death Dis* 6: e1768.

[80] Harris KG, Coyne CB (2015) Unc93b Induces Apoptotic Cell Death and Is Cleaved by Host and Enteroviral Proteases. *PLoS One* 10: e0141383.

[81] Tsoi H, Li L, Chen ZS, Lau KF, Tsui SK, et al. (2014) The SARS-coronavirus membrane protein induces apoptosis via interfering with PDK1-PKB/Akt signalling. *Biochem J* 464: 439-447.

[82] Ding L, Huang Y, Du Q, Dong F, Zhao X, et al. (2014) TGEV nucleocapsid protein induces cell cycle arrest and apoptosis through activation of p53 signaling. *Biochem Biophys Res Commun* 445: 497-503.

[83] Huang Y, Ding L, Li Z, Dai M, Zhao X, et al. (2013) Transmissible gastroenteritis virus infection induces cell apoptosis via activation of p53 signalling. *J Gen Virol* 94: 1807-1817.

[84] Fuentes-Gonzalez AM, Contreras-Paredes A, Manzo-Merino J, Lizano M (2013). The modulation of apoptosis by oncogenic viruses. *Virol J* 10: 182.

[85] Bargonetti J, Reynisdottir I, Friedman PN, Prives C (1992) Site-specific binding of wild-type p53 to cellular DNA is inhibited by SV40 T antigen and mutant p53. *Genes Dev* 6: 1886-1898.

[86] Yew PR, Liu X, Berk AJ (1994) Adenovirus E1B oncoprotein tethers a transcriptional repression domain to p53. *Genes Dev* 8: 190-202.

[87] Moore M, Horikoshi N, Shenk T (1996) Oncogenic potential of the adenovirus E4orf6 protein. *Proc Natl Acad Sci U S A* 93: 11295-11301.

[88] Scheffner M, Werness BA, Huibregtse JM, Levine AJ, Howley PM (1990). The E6 oncoprotein encoded by human papillomavirus types 16 and 18 promotes the degradation of p53. *Cell* 63: 1129-1136.

[89] Wang XW, Forrester K, Yeh H, Feitelson MA, Gu JR, et al. (1994) Hepatitis B virus X protein inhibits p53 sequence-specific DNA binding, transcriptional activity, and association with transcription factor ERCC3. *Proc Natl Acad Sci U S A* 91: 2230-2234.

[90] Lazo PA, Santos CR (2011) Interference with p53 functions in human viral infections, a target for novel antiviral strategies? *Rev Med Virol* 21: 285-300.

[91] Li Y, Zhang Q, Liu Y, Luo Z, Kang L, et al. (2012) Hepatitis C virus activates Bcl-2 and MMP-2 expression through multiple cellular signaling pathways. *J Virol* 86: 12531-12543.

[92] Alenzi FQ, El-Nashar EM, Al-Ghamdi SS, Abbas MY, Hamad AM, et al. (2010) Original Article: Investigation of Bcl-2 and PCNA in Hepatocellular Carcinoma: Relation to Chronic HCV. *J Egypt Natl Canc Inst* 22: 87-94.

[93] Chen G, Branton PE, Yang E, Korsmeyer SJ, Shore GC (1996) Adenovirus E1B 19-kDa death suppressor protein interacts with Bax but not with Bad. *J Biol Chem* 271: 24221-24225.

[94] Han J, Sabbatini P, Perez D, Rao L, Modha D, et al. (1996) The E1B 19K protein blocks apoptosis by interacting with and inhibiting the p53-inducible and death-promoting Bax protein. *Genes Dev* 10: 461-477.

[95] Smith CA (1995) A novel viral homologue of Bcl-2 and Ced-9. *Trends Cell Biol* 5: 344.

[96] Alibek K, Irving S, Sautbayeva Z, Kakpenova A, Bekmurzayeva A, et al. (2014) Disruption of Bcl-2 and Bcl-xL by viral proteins as a possible cause of cancer. *Infect Agent Cancer* 9: 44.

[97] Boya P, Roumier T, Andreau K, Gonzalez-Polo RA, Zamzami N, et al. (2003) Mitochondrion-targeted apoptosis regulators of viral origin. *Biochem Biophys Res Commun* 304: 575-581.

[98] Kincaid RP, Burke JM, Sullivan CS (2012) RNA virus microRNA that mimics a B-cell oncomiR. *Proc Natl Acad Sci U S A* 109: 3077-3082.

[99] Wang L, Li G, Yao ZQ, Moorman JP, Ning S (2015) MicroRNA regulation of viral immunity, latency, and carcinogenesis of selected tumor viruses and HIV. *Rev Med Virol* 25: 320-341.

[100] Gougeon ML, Montagnier L (1993) Apoptosis in AIDS. *Science* 260: 1269-1270.

[101] Garg H, Mohl J, Joshi A (2012) HIV-1 induced bystander apoptosis. Viruses 4: 3020-3043.

[102] Baloul L, Camelo S, Lafon M (2004) Up-regulation of Fas ligand (FasL) in the central nervous system: a mechanism of immune evasion by rabies virus. *J Neurovirol* 10: 372-382.

[103] Baloul L, Lafon M (2003) Apoptosis and rabies virus neuroinvasion. *Biochimie* 85: 777-788.

[104] Vuorinen T, Peri P, Vainionpaa R (2003) Measles virus induces apoptosis in uninfected bystander T cells and leads to granzyme B and caspase activation in peripheral blood mononuclear cell cultures. *Eur J Clin Invest* 33: 434-442.

[105] Skaletskaya A, Bartle LM, Chittenden T, McCormick AL, Mocarski ES, et al. (2001) A cytomegalovirus-encoded inhibitor of apoptosis that suppresses caspase-8 activation. *Proc Natl Acad Sci U S A* 98: 7829-7834.

[106] Hayajneh WA, Colberg-Poley AM, Skaletskaya A, Bartle LM, Lesperance MM, et al. (2001) The sequence and antiapoptotic functional domains of the human cytomegalovirus UL37 exon 1 immediate early protein are conserved in multiple primary strains. *Virology* 279: 233-240.

[107] Arnoult D, Bartle LM, Skaletskaya A, Poncet D, Zamzami N, et al. (2004) Cytomegalovirus cell death suppressor vMIA blocks Bax- but not Bak-mediated apoptosis by binding and sequestering Bax at mitochondria. *Proc Natl Acad Sci U S A* 101: 7988-7993.

[108] Terhune S, Torigoi E, Moorman N, Silva M, Qian Z, et al. (2007) Human cytomegalovirus UL38 protein blocks apoptosis. *J Virol* 81: 3109-3123.

[109] Wang HW, Sharp TV, Koumi A, Koentges G, Boshoff C (2002) Characterization of an anti-apoptotic glycoprotein encoded by Kaposi's sarcoma-associated herpesvirus which resembles a spliced variant of human survivin. *EMBO J* 21: 2602-2615.

[110] Brinkmann MM, Pietrek M, Dittrich-Breiholz O, Kracht M, Schulz TF (2007) Modulation of host gene expression by the K15 protein of Kaposi's sarcoma-associated herpesvirus. *J Virol* 81: 42-58.

[111] Cooray S, Bahar MW, Abrescia NG, McVey CE, Bartlett NW, et al. (2007) Functional and structural studies of the vaccinia virus virulence factor N1 reveal a Bcl-2-like anti-apoptotic protein. J Gen Virol 88: 1656-1666.

[112] Postigo A, Cross JR, Downward J, Way M (2006) Interaction of F1L with the BH3 domain of Bak is responsible for inhibiting vaccinia-induced apoptosis. *Cell Death Differ* 13: 1651-1662.

[113] Taylor JM, Quilty D, Banadyga L, Barry M (2006) The vaccinia virus protein F1L interacts with Bim and inhibits activation of the pro-apoptotic protein Bax. *J Biol Chem* 281: 39728-39739.

[114] Davies MV, Chang HW, Jacobs BL, Kaufman RJ (1993) The E3L and K3L vaccinia virus gene products stimulate translation through inhibition of the double-stranded RNA-dependent protein kinase by different mechanisms. *J Virol* 67: 1688-1692.

[115] Kettle S, Alcami A, Khanna A, Ehret R, Jassoy C, et al. (1997) Vaccinia virus serpin B13R (SPI-2) inhibits interleukin-1beta-converting enzyme and protects virus-infected cells from TNF- and Fas-mediated apoptosis, but does not prevent IL-1beta-induced fever. *J Gen Virol* 78 (Pt 3): 677-685.

[116] Tewari M, Telford WG, Miller RA, Dixit VM (1995) CrmA, a poxvirus-encoded serpin, inhibits cytotoxic T-lymphocyte-mediated apoptosis. *J Biol Chem* 270: 22705-22708.

[117] Hu FQ, Smith CA, Pickup DJ (1994) Cowpox virus contains two copies of an early gene encoding a soluble secreted form of the type II TNF receptor. *Virology* 204: 343-356.

[118] Chou J, Roizman B (1992) The gamma 1(34.5) gene of herpes simplex virus 1 precludes neuroblastoma cells from triggering total shutoff of protein synthesis characteristic of programed cell death in neuronal cells. *Proc Natl Acad Sci U S A* 89: 3266-3270.

[119] Munger J, Roizman B (2001) The US3 protein kinase of herpes simplex virus 1 mediates the posttranslational modification of BAD and prevents BAD-induced programmed cell death in the absence of other viral proteins. *Proc Natl Acad Sci U S A* 98: 10410-10415.

[120] Devergne O, Hatzivassiliou E, Izumi KM, Kaye KM, Kleijnen MF, et al. (1996) Association of TRAF1, TRAF2, and TRAF3 with an Epstein-Barr virus LMP1 domain important for B-lymphocyte transformation: role in NF-kappaB activation. *Mol Cell Biol* 16: 7098-7108.

[121] Kaye KM, Devergne O, Harada JN, Izumi KM, Yalamanchili R, et al. (1996) Tumor necrosis factor receptor associated factor 2 is a mediator of NF-kappa B activation by latent infection membrane protein 1, the Epstein-Barr virus transforming protein. *Proc Natl Acad Sci U S A* 93: 11085-11090.

[122] Silins SL, Sculley TB (1995) Burkitt's lymphoma cells are resistant to programmed cell death in the presence of the Epstein-Barr virus latent antigen EBNA-4. *Int J Cancer* 60: 65-72.

[123] Henderson S, Huen D, Rowe M, Dawson C, Johnson G, et al. (1993) Epstein-Barr virus-coded BHRF1 protein, a viral homologue of Bcl-2, protects human B cells from programmed cell death. *Proc Natl Acad Sci U S A* 90: 8479-8483.

[124] Perez D, White E (1998) E1B 19K inhibits Fas-mediated apoptosis through FADD-dependent sequestration of FLICE. *J Cell Biol* 141: 1255-1266.

[125] Teodoro JG, Branton PE (1997) Regulation of p53-dependent apoptosis, transcriptional repression, and cell transformation by phosphorylation of the 55-kilodalton E1B protein of human adenovirus type 5. *J Virol* 71: 3620-3627.

[126] Dahl J, Jurczak A, Cheng LA, Baker DC, Benjamin TL (1998) Evidence of a role for phosphatidylinositol 3-kinase activation in the blocking of apoptosis by polyomavirus middle T antigen. *J Virol* 72: 3221-3226.

[127] Kvansakul M, van Delft MF, Lee EF, Gulbis JM, Fairlie WD, et al. (2007) A structural viral mimic of prosurvival Bcl-2: a pivotal role for sequestering proapoptotic Bax and Bak. *Mol Cell* 25: 933-942.

[128] Messud-Petit F, Gelfi J, Delverdier M, Amardeilh MF, Py R, et al. (1998) Serp2, an inhibitor of the interleukin-1beta-converting enzyme, is critical in the pathobiology of myxoma virus. *J Virol* 72: 7830-7839.

[129] Cicin-Sain L, Ruzsics Z, Podlech J, Bubic I, Menard C, et al. (2008) Dominant-negative FADD rescues the in vivo fitness of a

cytomegalovirus lacking an antiapoptotic viral gene. *J Virol* 82: 2056-2064.

[130] Jurak I, Schumacher U, Simic H, Voigt S, Brune W (2008) Murine cytomegalovirus m38.5 protein inhibits Bax-mediated cell death. *J Virol* 82: 4812-4822.

[131] Cam M, Handke W, Picard-Maureau M, Brune W (2010) Cytomegaloviruses inhibit Bak- and Bax-mediated apoptosis with two separate viral proteins. *Cell Death Differ* 17: 655-665.

[132] Erdtmann L, Franck N, Lerat H, Le Seyec J, Gilot D, et al. (2003) The hepatitis C virus NS2 protein is an inhibitor of CIDE-B-induced apoptosis. *J Biol Chem* 278: 18256-18264.

[133] Lee SH, Kim YK, Kim CS, Seol SK, Kim J, et al. (2005) E2 of hepatitis C virus inhibits apoptosis. *J Immunol* 175: 8226-8235.

[134] Cruz CD, Palosaari H, Parisien JP, Devaux P, Cattaneo R, et al. (2006) Measles virus V protein inhibits p53 family member p73. *J Virol* 80: 5644-5650.

[135] Zauli G, Gibellini D, Milani D, Mazzoni M, Borgatti P, et al. (1993) Human immunodeficiency virus type 1 Tat protein protects lymphoid, epithelial, and neuronal cell lines from death by apoptosis. *Cancer Res* 53: 4481-4485.

[136] Wang XW, Gibson MK, Vermeulen W, Yeh H, Forrester K, et al. (1995) Abrogation of p53-induced apoptosis by the hepatitis B virus X gene. *Cancer Res* 55: 6012-6016.

[137] Denzler KL, Jacobs BL (1994) Site-directed mutagenic analysis of reovirus sigma 3 protein binding to dsRNA. *Virology* 204: 190-199.

[138] Lu Y, Wambach M, Katze MG, Krug RM (1995) Binding of the influenza virus NS1 protein to double-stranded RNA inhibits the activation of the protein kinase that phosphorylates the elF-2 translation initiation factor. *Virology* 214: 222-228.

[139] DeDiego ML, Nieto-Torres JL, Jimenez-Guardeno JM, Regla-Nava JA, Alvarez E, et al. (2011) Severe acute respiratory syndrome coronavirus envelope protein regulates cell stress response and apoptosis. *PLoS Pathog* 7: e1002315.

[140] Muhleisen A, Giaisi M, Kohler R, Krammer PH, Li-Weber M (2014) Tax contributes apoptosis resistance to HTLV-1-infected T cells via suppression of Bid and Bim expression. *Cell Death Dis* 5: e1575.

[141] Yuan L, Chen Z, Song S, Wang S, Tian C, et al. (2015) p53 degradation by a coronavirus papain-like protease suppresses type I interferon signaling. *J Biol Chem* 290: 3172-3182.

[142] Yuan J, Kroemer G (2010) Alternative cell death mechanisms in development and beyond. *Genes Dev* 24: 2592-2602.
[143] Christofferson DE, Yuan J (2010) Necroptosis as an alternative form of programmed cell death. *Curr Opin Cell Biol* 22: 263-268.
[144] Feoktistova M, Geserick P, Panayotova-Dimitrova D, Leverkus M (2012) Pick your poison: The Ripoptosome, a cell death platform regulating apoptosis and necroptosis. *Cell Cycle* 11: 460-467.
[145] Kaiser WJ, Sridharan H, Huang C, Mandal P, Upton JW, et al. (2013) Toll-like receptor 3-mediated necrosis via TRIF, RIP3, and MLKL. *J Biol Chem* 288: 31268-31279.
[146] Upton JW, Kaiser WJ, Mocarski ES (2012) DAI/ZBP1/DLM-1 complexes with RIP3 to mediate virus-induced programmed necrosis that is targeted by murine cytomegalovirus vIRA. *Cell Host Microbe* 11: 290-297.
[147] Thapa RJ, Nogusa S, Chen P, Maki JL, Lerro A, et al. (2013) Interferon-induced RIP1/RIP3-mediated necrosis requires PKR and is licensed by FADD and caspases. *Proc Natl Acad Sci U S A* 110: E3109-3118.
[148] Dunai Z, Bauer PI, Mihalik R (2011) Necroptosis: biochemical, physiological and pathological aspects. Pathol Oncol Res 17: 791-800.
[149] Christofferson DE, Li Y, Hitomi J, Zhou W, Upperman C, et al. (2012) A novel role for RIP1 kinase in mediating TNFalpha production. *Cell Death Dis* 3: e320.
[150] Cho YS, Challa S, Moquin D, Genga R, Ray TD, et al. (2009) Phosphorylation-driven assembly of the RIP1-RIP3 complex regulates programmed necrosis and virus-induced inflammation. *Cell* 137: 1112-1123.
[151] Zhang YY, Liu H (2013) Connections Between Various Trigger Factors and the RIP1/ RIP3 Signaling Pathway Involved in Necroptosis. *Asian Pac J Cancer Prev* 14: 7069-7074.
[152] Lalaoui N, Lindqvist LM, Sandow JJ, Ekert PG (2015) The molecular relationships between apoptosis, autophagy and necroptosis. *Semin Cell Dev Biol* 39: 63-69.
[153] Li J, McQuade T, Siemer AB, Napetschnig J, Moriwaki K, et al. (2012) The RIP1/RIP3 necrosome forms a functional amyloid signaling complex required for programmed necrosis. *Cell* 150: 339-350.
[154] Sun L, Wang H, Wang Z, He S, Chen S, et al. (2012) Mixed lineage kinase domain-like protein mediates necrosis signaling downstream of RIP3 kinase. *Cell* 148: 213-227.

[155] Cai Z, Jitkaew S, Zhao J, Chiang HC, Choksi S, et al. (2014) Plasma membrane translocation of trimerized MLKL protein is required for TNF-induced necroptosis. *Nat Cell Biol* 16: 55-65.

[156] Chen X, Li W, Ren J, Huang D, He WT, et al. (2014) Translocation of mixed lineage kinase domain-like protein to plasma membrane leads to necrotic cell death. *Cell Res* 24: 105-121.

[157] Marshall KD, Baines CP (2014) Necroptosis: is there a role for mitochondria? *Front Physiol* 5: 323.

[158] Baines CP (2010) Role of the mitochondrion in programmed necrosis. *Front Physiol* 1: 156.

[159] Galluzzi L, Kepp O, Krautwald S, Kroemer G, Linkermann A (2014) Molecular mechanisms of regulated necrosis. *Semin Cell Dev Biol* 35: 24-32.

[160] Sridharan H, Upton JW (2014) Programmed necrosis in microbial pathogenesis. *Trends Microbiol* 22: 199-207.

[161] Kaiser WJ, Upton JW, Mocarski ES (2013) Viral modulation of programmed necrosis. *Curr Opin Virol* 3: 296-306.

[162] Dobbelstein M, Shenk T (1996) Protection against apoptosis by the vaccinia virus SPI-2 (B13R) gene product. *J Virol* 70: 6479-6485.

[163] Chan FK, Shisler J, Bixby JG, Felices M, Zheng L, et al. (2003) A role for tumor necrosis factor receptor-2 and receptor-interacting protein in programmed necrosis and antiviral responses. J Biol Chem 278: 51613-51621.

[164] Fliss PM, Brune W (2012) Prevention of cellular suicide by cytomegaloviruses. *Viruses* 4: 1928-1949.

[165] Upton JW, Kaiser WJ, Mocarski ES (2010) Virus inhibition of RIP3-dependent necrosis. *Cell Host Microbe* 7: 302-313.

[166] Clarke P, Meintzer SM, Gibson S, Widmann C, Garrington TP, et al. (2000) Reovirus-induced apoptosis is mediated by TRAIL. *J Virol* 74: 8135-8139.

[167] Clarke P, Richardson-Burns SM, DeBiasi RL, Tyler KL (2005) Mechanisms of apoptosis during reovirus infection. *Curr Top Microbiol Immunol* 289: 1-24.

[168] Berger AK, Danthi P (2013) Reovirus activates a caspase-independent cell death pathway. *MBio* 4: e00178-00113.

[169] Yatim N, Albert ML (2011) Dying to replicate: the orchestration of the viral life cycle, cell death pathways, and immunity. *Immunity* 35: 478-490.

[170] Rodrigue-Gervais IG, Labbe K, Dagenais M, Dupaul-Chicoine J, Champagne C, et al. (2014) Cellular inhibitor of apoptosis protein cIAP2 protects against pulmonary tissue necrosis during influenza virus infection to promote host survival. *Cell Host Microbe* 15: 23-35.

[171] Pan T, Wu S, He X, Luo H, Zhang Y, et al. (2014) Necroptosis takes place in human immunodeficiency virus type-1 (HIV-1)-infected CD4+ T lymphocytes. *PLoS One* 9: e93944.

[172] Pinard M, Le Coupanec A, Desforges M, Talbot PJ (2015) Pivotal role of RIP1-dependent necroptosis in neuronal cell death induced by the human neuroinvasive coronavirus OC43.

[173] Guo H, Omoto S, Harris PA, Finger JN, Bertin J, et al. (2015) Herpes simplex virus suppresses necroptosis in human cells. *Cell Host Microbe* 17: 243-251.

[174] Omoto S, Guo H, Talekar GR, Roback L, Kaiser WJ, et al. (2015) Suppression of RIP3-dependent necroptosis by human cytomegalovirus. *J Biol Chem* 290: 11635-11648.

[175] Kaczmarek A, Vandenabeele P, Krysko DV (2013) Necroptosis: the release of damage-associated molecular patterns and its physiological relevance. *Immunity* 38: 209-223.

[176] Chien H, Dix RD (2012) Evidence for multiple cell death pathways during development of experimental cytomegalovirus retinitis in mice with retrovirus-induced immunosuppression: apoptosis, necroptosis, and pyroptosis. *J Virol* 86: 10961-10978.

[177] Son KN, Lipton HL (2015) Inhibition of Theiler's virus-induced apoptosis in infected murine macrophages results in necroptosis. *Virus Res* 195: 177-182.

[178] Huang Z, Wu SQ, Liang Y, Zhou X, Chen W, et al. (2015) RIP1/RIP3 binding to HSV-1 ICP6 initiates necroptosis to restrict virus propagation in mice. *Cell Host Microbe* 17: 229-242.

[179] Wang X, Li Y, Liu S, Yu X, Li L, et al. (2014) Direct activation of RIP3/MLKL-dependent necrosis by herpes simplex virus 1 (HSV-1) protein ICP6 triggers host antiviral defense. *Proc Natl Acad Sci U S A* 111: 15438-15443.

[180] Upton JW, Kaiser WJ, Mocarski ES (2008) Cytomegalovirus M45 cell death suppression requires receptor-interacting protein (RIP) homotypic interaction motif (RHIM)-dependent interaction with RIP1. *J Biol Chem* 283: 16966-16970.

[181] Harris KG, Morosky SA, Drummond CG, Patel M, Kim C, et al. (2015) RIP3 Regulates Autophagy and Promotes Coxsackievirus B3 Infection of Intestinal Epithelial Cells. *Cell Host Microbe* 18: 221-232.

[182] Virgin HW, Wherry EJ, Ahmed R (2009) Redefining chronic viral infection. *Cell* 138: 30-50.

[183] Griffin DE, Hardwick JM (1997) Regulators of apoptosis on the road to persistent alphavirus infection. *Annu Rev Microbiol* 51: 565-592.

[184] Ghosh Roy S, Sadigh B, Datan E, Lockshin RA, Zakeri Z (2014) Regulation of cell survival and death during Flavivirus infections. *World J Biol Chem* 5: 93-105.

[185] Oldstone MB (2006) Viral persistence: parameters, mechanisms and future predictions. *Virology* 344: 111-118.

[186] Huang C, Zhang Q, Feng WH (2015) Regulation and evasion of antiviral immune responses by porcine reproductive and respiratory syndrome virus. *Virus Res* 202: 101-111.

[187] Ahmed R, Morrison L, Knipe D (1997) Viral persistence. In: Nathanson N, al e, editors. Viral pathogenesis. Philadelphia: *Lippincott Raven.* pp. 181-205.

[188] Miller SD, Vanderlugt CL, Begolka WS, Pao W, Yauch RL, et al. (1997) Persistent infection with Theiler's virus leads to CNS autoimmunity via epitope spreading. *Nat Med* 3: 1133-1136.

[189] Oldstone MB (1989) Viral persistence. *Cell* 56: 517-520.

[190] de Groot R. J., Baker, S. C., Baric, R., Enjuanes, L., Gorbalenya, A. E., Holmes, K. V., Perlman, S. Poon, L., Rottier, P. J. M., Talbot, P. J., Woo, P. C. Y., and Ziebuhr, J., editor (2012) Family Coronaviridae. New York: *Elsevier.* 806-827 p.

[191] Weiss SRaL, J. L. (2011) Coronavirus pathogenesis. In: Maramorosch K, Shatkin, A. J., Murphy, F. A., editor. Advances in Virus Research. Boston: *Elsevier.* pp. 85-164.

[192] Castro RF, Evans GD, Jaszewski A, Perlman S (1994) Coronavirus-induced demyelination occurs in the presence of virus-specific cytotoxic T cells. *Virology* 200: 733-743.

[193] Houtman JJ, Fleming JO (1996) Pathogenesis of mouse hepatitis virus-induced demyelination. *J Neurovirol* 2: 361-376.

[194] Wege H, Stuhler A, Lassmann H (1998) Coronavirus infection and demyelination. Sequence conservation of the S-gene during persistent infection of Lewis-rats. *Adv Exp Med Biol* 440: 767-773.

[195] Buchmeier MJ, Lane TE (1999) Viral-induced neurodegenerative disease. *Curr Opin Microbiol* 2: 398-402.

[196] Stohlman SA, Hinton DR (2001) Viral induced demyelination. *Brain Pathol* 11: 92-106.

[197] Lavi E, Suzumura A, Hirayama M, Highkin MK, Dambach DM, et al. (1987) Coronavirus mouse hepatitis virus (MHV)-A59 causes a persistent, productive infection in primary glial cell cultures. *Microb Pathog* 3: 79-86.

[198] Chen W, Baric RS (1996) Molecular anatomy of mouse hepatitis virus persistence: coevolution of increased host cell resistance and virus virulence. *J Virol* 70: 3947-3960.

[199] Thorley-Lawson DA, Hawkins JB, Tracy SI, Shapiro M (2013) The pathogenesis of Epstein-Barr virus persistent infection. *Curr Opin Virol* 3: 227-232.

[200] Mitchell JK, Lemon SM, McGivern DR (2015) How do persistent infections with hepatitis C virus cause liver cancer? *Curr Opin Virol* 14: 101-108.

[201] Bartsch H, Nair J (2006) Chronic inflammation and oxidative stress in the genesis and perpetuation of cancer: role of lipid peroxidation, DNA damage, and repair. *Langenbecks Arch Surg* 391: 499-510.

[202] Herzer K, Sprinzl MF, Galle PR (2007) Hepatitis viruses: live and let die. *Liver Int* 27: 293-301.

[203] Nakamoto Y, Kaneko S (2003) Mechanisms of viral hepatitis induced liver injury. *Curr Mol Med* 3: 537-544.

[204] Dupont L, Reeves MB (2015) Cytomegalovirus latency and reactivation: recent insights into an age old problem. *Rev Med Virol.*

[205] Sinclair J (2008) Human cytomegalovirus: Latency and reactivation in the myeloid lineage. *J Clin Virol* 41: 180-185.

[206] Griffiths P, Baraniak I, Reeves M (2015) The pathogenesis of human cytomegalovirus. *J Pathol* 235: 288-297.

[207] Talbot PJ, Arnold D, Antel JP (2001) Virus-induced autoimmune reactions in the CNS. *Current Topics in Microbiology and Immunology* 253: 247-271.

[208] de la Torre JC, Mallory M, Brot M, Gold L, Koob G, et al. (1996) Viral persistence in neurons alters synaptic plasticity and cognitive functions without destruction of brain cells. *Virology* 220: 508-515.

[209] Big C, Reineck LA, Aronoff DM (2009) Viral infections of the central nervous system: a case-based review. *Clin Med Res* 7: 142-146.

[210] Giraudon P, Bernard A (2010) Inflammation in neuroviral diseases. *J Neural Transm* 117: 899-906.

[211] Whitley RJ, Gnann JW (2002) Viral encephalitis: familiar infections and emerging pathogens. *Lancet 359*: 507-513.

[212] Hankins DG, Rosekrans JA (2004) Overview, prevention, and treatment of rabies. *Mayo Clinic proceedings Mayo Clinic* 79: 671-676.

[213] Aurelian L (2005) HSV-induced apoptosis in herpes encephalitis. *Current Topics in Microbiology and Immunology* 289: 79-111.

[214] Mackenzie JS, Gubler DJ, Petersen LR (2004) Emerging flaviviruses: the spread and resurgence of Japanese encephalitis, West Nile and dengue viruses. *Nature Medicine* 10: S98-109.

[215] Sips GJ, Wilschut J, Smit JM (2012) Neuroinvasive flavivirus infections. *Rev Med Virol* 22: 69-87.

[216] Neal JW (2014) Flaviviruses are neurotropic, but how do they invade the CNS? *J Infect* 69: 203-215.

[217] Mueller S, Wimmer E, Cello J (2005) Poliovirus and poliomyelitis: a tale of guts, brains, and an accidental event. *Virus Research* 111: 175-193.

[218] Rhoades RE, Tabor-Godwin JM, Tsueng G, Feuer R (2011) Enterovirus infections of the central nervous system. *Virology* 411: 288-305.

[219] Arbour N, Day R, Newcombe J, Talbot PJ (2000) Neuroinvasion by human respiratory coronaviruses. *J Virol* 74: 8913-8921.

[220] Desforges M, Le Coupanec A, Stodola JK, Meessen-Pinard M, Talbot PJ (2014) Human coronaviruses: viral and cellular factors involved in neuroinvasiveness and neuropathogenesis. *Virus Res* 194: 145-158.

[221] Morfopoulo S, J. R. B, E. G. D, G. A, A. V, et al. (2015) Coronavirus HCoV-OC43 associated with fatal encephalitis.

[222] Mattson MP, Haughey NJ, Nath A (2005) Cell death in HIV dementia. *Cell Death and Differentiation* 12 Suppl 1: 893-904.

[223] Nath A, Berger J (2004) HIV Dementia. Current Treatment Options in Neurology 6: 139-151.

[224] Gordon J, Gallia GL, Del Valle L, Amini S, Khalili K (2000) Human polyomavirus JCV and expression of myelin genes. *Journal of Neurovirology* 6 Suppl 2: S92-97.

[225] Shishido-Hara Y (2015) Progressive multifocal leukoencephalopathy: Dot-shaped inclusions and virus-host interactions. *Neuropathology* 35: 487-496.

[226] Weissert R (2011) Progressive multifocal leukoencephalopathy. *Journal of Neuroimmunology* 231: 73-77.

[227] Kaplan JE, Osame M, Kubota H, Igata A, Nishitani H, et al. (1990) The risk of development of HTLV-I-associated myelopathy/tropical spastic

paraparesis among persons infected with HTLV-I. *J Acquir Immune Defic Syndr* 3: 1096-1101.

[228] Mortreux F, Gabet AS, Wattel E (2003) Molecular and cellular aspects of HTLV-1 associated leukemogenesis in vivo. *Leukemia* 17: 26-38.

[229] Itzhaki RF, Wozniak MA, Appelt DM, Balin BJ (2004) Infiltration of the brain by pathogens causes Alzheimer's disease. *Neurobiology of Aging* 25: 619-627.

[230] Arbour N, Day R, Newcombe J, Talbot PJ (2000) Neuroinvasion by human respiratory coronaviruses. *Journal of Virology* 74: 8913-8921.

[231] Arbour N, Cote G, Lachance C, Tardieu M, Cashman NR, et al. (1999) Acute and persistent infection of human neural cell lines by human coronavirus OC43. *Journal of Virology* 73: 3338-3350.

[232] Arbour N, Ekande S, Cote G, Lachance C, Chagnon F, et al. (1999) Persistent infection of human oligodendrocytic and neuroglial cell lines by human coronavirus 229E. *Journal of Virology* 73: 3326-3337.

[233] Jacomy H, Fragoso G, Almazan G, Mushynski WE, Talbot PJ (2006) Human coronavirus OC43 infection induces chronic encephalitis leading to disabilities in BALB/C mice. *Virology* 349: 335-346.

[234] Jacomy H, St-Jean JR, Brison E, Marceau G, Desforges M, et al. (2010) Mutations in the spike glycoprotein of human coronavirus OC43 modulate disease in BALB/c mice from encephalitis to flaccid paralysis and demyelination. *J Neurovirol* 16: 279-293.

[235] Brison E, Jacomy H, Desforges M, Talbot PJ (2011) Glutamate excitotoxicity is involved in the induction of paralysis in mice after infection by a human coronavirus with a single point mutation in its spike protein. *J Virol* 85: 12464-12473.

In: Programmed Cell Death in Plants and Animals ISBN: 978-1-63484-505-2
Editor: Josephine Rice 

*Chapter 3*

# CONTROL OF LEAF CELL DEATH BY REACTIVE OXYGEN SPECIES

***Bartolomé Sabater and Mercedes Martín***
Department of Life Sciences (Plant Physiology)
University of Alcalá. Spain

## ABSTRACT

The leaf is a complex organ composed of different tissues and cells that are subjected to a variety of environmental factors affecting the course of its aging, senescence and death. Most investigations on leaf death have been focused on the senescence of photosynthetic mesophyll cells, whose characteristic loss of chlorophyll becomes the reference symptom preceding leaf death. Autumnal and monocarpic leaf deaths are developmental processes genetically programmed but, at the cell level, many changes that precede them are similar to those of the senescence of shaded basal leaves and even to those of the stress-dependent leaf aging. Most knowledge of the molecular processes in programmed leaf death has been gained through the investigation with the last two deaths, provoked by developmental and environmental factors that in strict sense are not programmed leaf death. However, analogies with processes in programmed cell death in animals provide unifying views where reactive oxygen species (ROS, mainly singlet oxygen $^1O_2$, anion radical superoxide $O_2^{\bullet-}$ and hydrogen peroxide $H_2O_2$) emerge as key intracellular signals.

The steady state level of each ROS depends on generation and scavenging activities closely linked, respectively, to the photosynthetic electron transport and to systems that protect cells from the destructive action of ROS and, frequently, the destructive actions of ROS are hardly distinguishable from their signaling role in cell death. By controlling the levels of different ROS, chloroplasts regulate leaf death and assume roles similar to several roles attributed to mitochondria in programmed animal cells. High ratios ($^1O_2$ plus $O_2\bullet^-$)/$H_2O_2$ seem key to shoot the autocatalytic cell path to death where hormones and transcriptional factors join to change the feed-back regulation of cell processes to a feed-forward spiral of deleterious reactions. In this chapter, we will discuss the control of the levels of different ROS and their interaction with hormones and transcriptional factors in relation with programmed cell death in leaves.

## INTRODUCTION

Except for a few plants where basal meristem persists, the leaf is an organ with limited growth where, when it reaches the characteristic form and size of adult stage, all its meristematic cells have differentiated into a variety of tissues among which mesophyll photosynthetic are the majority. Lacking meristematic cells, the life of an adult leaf is time-limited and spans from a few days in many herbaceous to several years in some conifers. Chlorophyll loss is the most characteristic symptom of leaf senescence preceding its death and it is one feature of the dismantling of chloroplast machinery.

Leaf death may be one aspect of the whole plant death, as in monocarpic, or it may occur while the whole plant remains alive, as in the autumnal leaf death of many perennials. Both, monocarpic and autumnal leaf deaths are developmental processes genetically programmed: only slightly (and frequently indirectly) affected by environmental factors. In addition, there are processes of leaf senescence and death, strongly affected by environmental factors, such as light intensity, temperature or nutrient availability. This is the case of the basal leaves of rapidly growing gramineans, whose senescence and death are linked to the simultaneous development of other organs of the plants. In both monocarpic and basal leaf, death is frequently delayed by scission of, respectively, reproductive buds and growing young leaves that, otherwise, act as sinks of the degradation products of senescent leaves. Finally, a leaf may die by a prolonged accumulation of hazardous damages, a process named aging, or by a burst of damages under biotic or abiotic stresses.

Except for the rapid death under biotic or abiotic stresses, leaf senescence preceding death is an organized process through which most valuable components of the leaves, as nitrogen present in proteins, are mobilized to young, storage or reproduction organs. This mobilization requires the ordered expression of genes encoding enzymatic activities that degrade, among others, chloroplast proteins (the main fraction of leaf protein, Dalling 1985) and reconvert amino acid nitrogen to exportable forms as glutamine and asparagine (Morita 1980; Feller and Fischer 1994). The commitment of plants to recover valuable nutrients (as nitrogen, phosphorus, potassium or magnesium) in seeds and growing young organs justifies the evolutionary selection of leaf senescence traits: ordered and efficient in dismantling cell structures and sending out components in coordination with the development of other organs in the plant.

The amount of Reactive Oxygen Species (ROS) increase in all types of leaf senescence and in aging, with respect to young and adult healthy leaves. ROS are mainly produced through side reactions in the photosynthetic machinery of chloroplasts (Apel and Hirt 2004; Sabater and Martín 2013a,b). ROS are highly reactive damaging agents that could demolish cell structures. In addition, ROS are signals that control the expression of genes that help cells to survive under different stress conditions or, alternatively, to lead them to death. Therefore, through ROS production and scavenging, chloroplasts have an active role controlling leaf senescence (Zapata et al. 2005; Chen et al. 2010; Doyle et al. 2010; Van Doorn and Yoshimoto 2010; Sabater and Martín 2013a,b). Probably, the damaging effect of ROS is a side effect that does not interfere with the organized processes of cell death and evidences suggest that the relative levels of the different ROS determine the expression of genes for either live or suicide metabolism of leaf cell (Gechev et al. 2006). Among genes expressed for cell death are those encoding enzymes required for the mobilization of valuable leaf components and, in general, for the orchestration of programmed cell death during senescence (SAG, senescence associated genes): an ordered process hardly compatible with the unspecific hazardous damage effects of ROS.

When compared with other cellular signals, the investigation of the role of ROS poses specific challenges related to their generation and scavenging, their translocation among cell compartments and their interactions with other cellular signals. In recent years, improved quantification of specific ROS and indirect evidences based on changes of activities favoring formation or scavenging specific ROS give insights on the involvement of each specific ROS in programmed death of leaf cells. On the basis of their low half-life and permeability across membranes the primary action of the ROS singlet oxygen, $^1O_2$, and anion radical superoxide, $O_2\bullet^-$, must occur in the same compartment where they are formed, mainly chloroplasts. The other key ROS, hydrogen peroxide, $H_2O_2$, is by far the most stable and it is formed in, essentially, all cell compartments, although probably chloroplast is the main source of $H_2O_2$ in all stages of leaf development. $H_2O_2$ freely crosses membranes and its primary action may occur in cell compartments different from those generating it. There are several proposals for the primary action of the different ROS, that is: for their direct effect generating or modifying other signals in the complex network that regulate cell development, including cell death.

## GENERATION AND SCAVENGING OF ROS IN CHLOROPLASTS

In leaves, ROS are mainly generated in chloroplasts that are also the main places that scavenge $^1O_2$ and $O_2\bullet^-$. The contribution of different compartments to scavenging $H_2O_2$ is uncertain. $H_2O_2$ may be consumed by different peroxidases (PXs) present in several compartments (among them chloroplasts) that, having high affinity for $H_2O_2$ ($K_m$ in the range 0.1 to 1 mM), should maintain its low concentration, as long as there is enough supply of electron donors: the other peroxidase substrate, which varies among the different peroxidases. Catalase enzyme decomposes $H_2O_2$, it is also present in several compartments (but not in chloroplasts) and does not require a second substrate donor of electron. However, the affinity of catalase for $H_2O_2$ is low ($K_m$ 10 to 100 mM) and can hardly decrease the concentration of $H_2O_2$ below 1 mM.

To minimize the damaging effects of ROS, the scavenging machineries maintain low steady state concentrations of ROS, compatible with high photosynthetic activity under widely variable and frequently stressing environments. Therefore, levels of $^1O_2$, $O_2\bullet^-$ and $H_2O_2$ that control cell death are probably low too and their damaging effects do not significantly contribute to the dismantling cell components, at least during early stages of leaf senescence.

The relative level of each ROS depends on relative levels of the diverse activities involved in the generation and the scavenging of the different ROS. Generation of ROS is closely related to activities of the photosynthetic electron transport in chloroplasts. Figure 1 shows the main processes generating $^1O_2$, $O_2\bullet^-$ and $H_2O_2$ in chloroplasts of illuminated leaves. Most $^1O_2$

is formed in photosystem II (PSII) by transfer of electronic excitation from chlorophyll excited at the triplet state ($^3Chl^*$) to molecular oxygen $O_2$. $^3Chl^*$ is formed from chlorophyll excited at the singlet state (Chl*). Accordingly, the fraction of chlorophyll $^3Chl^*$ in chloroplasts increases when that of Chl* rises because the absorption of light exceeds the capacity to drain electrons from photosynthetic electron transporters, mainly by assimilation of $CO_2$. Most $O_2{\bullet}^-$ is formed in photosystem I (PSI) by one electron transfer from $A/Fd_{red}$ (reduced non-heme iron-sulfur proteins, including ferredoxin) to molecular oxygen $O_2$ (Mehler reaction). $O_2{\bullet}^-$ may be also generated in a lesser amount by electron transfer from reduced non-heme iron-sulfur components of the thylakoid Ndh complex (Ndh) and by interaction of $^1O_2$ with different organic and inorganic molecules (not represented in Figure 1). $H_2O_2$ is mainly generated from $O_2{\bullet}^-$ by the reaction: $2\ O_2{\bullet}^- + 2\ H^+ \rightarrow H_2O_2 + O_2$, catalyzed by superoxide dismutase (SOD), which also is the main way for $O_2{\bullet}^-$ degradation. As pointed, $H_2O_2$ is consumed in chloroplasts when it oxidizes different substrates in reactions catalyzed by peroxidases.

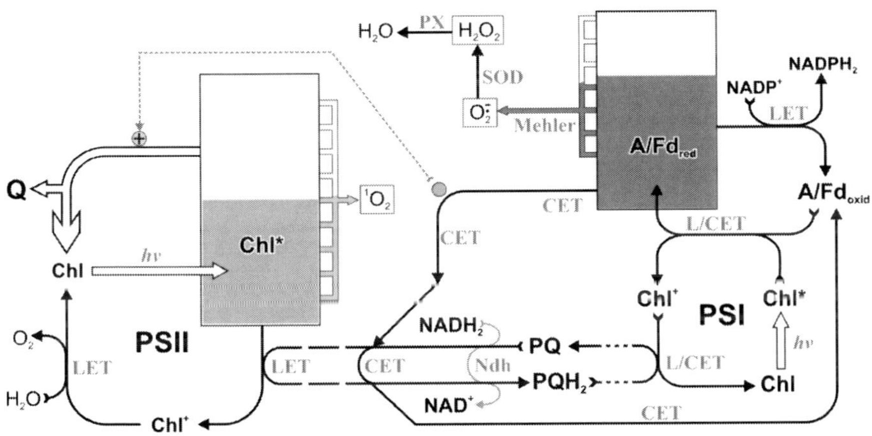


Figure 1. Generation of ROS in chloroplasts. Non-conventional representation of the photosynthetic electron transport to highlight processes generating ROS. Electron transport from $H_2O$ to $NADP^+$ is depicted by arrows marked with LET (linear electron transport) connecting water oxidation (down left), photosystem II (**PSII**), photosystem I (**PSI**) and the final $NADP^+$ reduction step (up right). Specific steps of cyclic electron transport around **PSI** are marked as CET and common steps of linear and cyclic electron transport are marked as L/CET. A/Fd, non-heme-iron sulfur proteins (including ferredoxin); Ndh, the thylakoid NADH dehydrogenase complex; PX, peroxidase; **Q**, energy dissipated as heat. SOD, superoxide dismutase. Other lettering and symbols are described in the text.

Levels of Chl* in PSII and of A/$Fd_{red}$ in PSI (represented in Figure 1 by partially filled tanks with vertically disposed leakage exits) play key roles for the generation of, respectively, $^1O_2$ and $O_2{\bullet}^-$. The higher levels of Chl* and A/$Fd_{red}$, higher are the rates of generation of, respectively, $^1O_2$ and $O_2{\bullet}^-$. The level of Chl* in PSII increases with the intensity of light, which excites Chl to Chl* by absorption of one photon ($h\nu$), and decreases by Chl* oxidation to $Chl^+$ (loss of the excited electron, which is transferred to plastoquinone PQ, A/$Fd_{red}$ and $NADP^+$ through different steps of the linear electron transport LET chain), by emission of fluorescent light, and by heat dissipation (Q) after transfer of excitation to zeaxanthin formed in the xanthophyll cycle (Eskling et al. 2001). The last process is stimulated by high light intensity and cyclic electron transport (CET) through thylakoid polarization. The rate of electron transfer from Chl* of PSII through LET essentially equals its use in the reduction of $CO_2$ (through NADPH) plus, to a much lower extent, the reduction of nitrate and the Mehler reaction. When light intensity exceeds the capacity of the Benson-Calvin cycle to drain electrons, Chl* and A/$Fd_{red}$ increase and the production of $^1O_2$ and $O_2{\bullet}^-$ rise. In these conditions, subsequent photoinhibition of PSII and activation of heat dissipation Q contribute to alleviate the generation of $^1O_2$ and $O_2{\bullet}^-$. In addition, increase of the SOD activity contributes to maintain low levels of $O_2{\bullet}^-$.

Therefore, Mehler reaction, SOD and PX activities are key players to maintain low levels of $O_2{\bullet}^-$. They also affect indirectly the level of $^1O_2$. In effect, as pointed out above, heat dissipation requires CET to maintain the polarized thylakoid membrane. When the absorption of light exceeds the electron drain toward the assimilation of $CO_2$, the ratio: reduced forms to oxidized forms of transporters of LET and CET becomes high, which decreases the rate of CET whose maximum requires balanced (poised) levels of reduced and oxidized forms of transporters. Low rate of CET, by over-reduction of transporters, decreases heat dissipation of the Chl* of PSII as indicated and, hence, facilitates over-production of $^1O_2$. Therefore, generation of $O_2{\bullet}^-$ by Mehler reaction alleviates over-reduction of transporters and restores maximum rates of CET allowing efficient heat dissipation of the excess of light energy and low production of $^1O_2$. In this way, the production of $O_2{\bullet}^-$ by Mehler reaction decreases the production of the more dangerous $^1O_2$. SOD and PX are necessary to quench the level of excess $O_2{\bullet}^-$ and $H_2O_2$ produced. Among PXs, plastoquinol peroxidase oxidizes reduced PQ with $H_2O_2$ (Zapata et al. 1998; Casano et al. 2000). Therefore, it consumes $H_2O_2$ and drains electrons, contributing to alleviate over-reduction of the photosynthetic electron transport chain.

Several stressing conditions, transitorily or long-term, increase the excess of light absorption over the capacity to drain electrons from the photosynthetic electron transport chain. The sudden increase of light intensity is the most obvious. In addition, lowering temperature (which decreases the rate of enzymes of the Benson-Calvin cycle) and water deficit (which closes stomata and decreases the availability of $CO_2$) are frequent stresses where maintaining efficient CET transport (Segura and Quiles 2015) is correlated with the control of levels of different ROS.

Low rate of the CET can also be produced by the opposite condition: a low ratio of the reduced to the oxidized forms of transporters. This is a frequent circumstance under rapidly fluctuating light intensities (sun fleck). During a period under high light intensity, photoinhibiton of PSII and active heat dissipation depresses the capacity of PSII to send electrons to LET. Depression become dramatic when light intensity suddenly and strongly decreases. Then, transporters would become over-oxidized and the rate of CET and polarization of the thylakoid membrane lowers. Recovery from photoinhibition and decrease of dissipating zeaxanthin may require between several seconds and minutes: too long for efficient photosynthesis, especially at high concentrations of $CO_2$ (Martín et al. 2015). To restore the balanced redox state of transporters and the consequently high rate of CET, the transitory low supply of electrons from PSII is supplemented by the thylakoid Ndh complex (see Figure 1) that transfers electrons from NADH to PQ. Therefore, under usual fluctuating lights, leaves meet stresses of alternating high and low light intensities where the photosynthetic electron transport chain alternates steps of electron drain by Mehler reaction, SOD and PX with steps of electron filling by the Ndh complex. Jointly, although not simultaneously, the successive reactions conform the chlororespiratory electron transport chain (Casano et al. 2000; Joët et al. 2002; Nixon and Rich 2007):

$$2\ NADH + 2\ H^+ + 2\ PQ \xrightarrow{\text{(Ndh complex)}} 2\ NAD^+ + 2\ PQH_2$$

$$2\ A/Fd_{oxid} + PQH_2 \xrightarrow{\text{LET}} 2\ A/Fd_{red} + PQ + 2H^+$$

$$2\ A/Fd_{red} + 2\ O_2 \xrightarrow{\text{Mehler reaction}} 2\ A/Fd_{oxid} + 2\ O_2{\bullet}^-$$

$$2\,O_2^{\bullet -} + 2\,H^+ \xrightarrow{\text{SOD}} O_2 + H_2O_2$$

$$H_2O_2 + PQH_2 \xrightarrow{\text{PX}} 2\,H_2O + PQ$$

Which results in the global reaction:

$$2\,NADH + 2\,H^+ + O_2 \longrightarrow 2\,NAD^+ + 2\,H_2O$$

that transports electrons from NADH to oxygen to maintain the poised redox level of electron transporters and, hence, appropriate rates of CET.

Accordingly, levels and activities of the chloroplast Ndh complex, SOD and PX increase under a variety of stresses (Bowler et al. 1992; Casano et al. 1994; Martín et al.1996; Casano et al.1999, 2001; Martín et al. 2004; Serrot et al. 2008; Paredes and Quiles 2013; Segura and Quiles 2015).

## SWITCHING OF ROS METABOLISM TO LEAF SENESCENCE

Appropriate balance of activities maintains low levels of $^1O_2$, $O_2^{\bullet -}$ and $H_2O_2$ to cope with different stress conditions, reduces damage by ROS to a minimum and induces defensive adaptation of the leaves against stressing agents. Several evidences point specifically to $H_2O_2$ as a signal in the induction of the defense responses against stress (Mubarakshina et al. 2010; Wiciarz et al. 2015). Certainly, longer half-life, mobility in the cell and lower toxicity make $H_2O_2$ more appropriate than $^1O_2$ and $O_2^{\bullet -}$ to address adaptation-surviving responses. The situation must be different if leaf survival does not yet provide evolutionary advantages for the plant. In other words: if the leaf should senesce and die.

In effect, although the Ndh complex and PX increases similarly under stress along leaf development since birth to senescence, the induction of chloroplastic SODs (Cu/Zn and Fe SODs) by photo-oxidative stress is impaired early during leaf senescence (Casano et al. 1994; Kurepa et al. 1997; Casano et al. 1999; Abarca et al. 2001a,b; Prochazkova et al. 2001; Ohe et al. 2005). The first consequence is an increase of $O_2^{\bullet -}$ and a decrease of $H_2O_2$ early in senescence. Further increases of poorly scavenged $O_2^{\bullet -}$ occur due to enhanced Mehler reactions at the level of non-heme iron-sulfur proteins of the increased Ndh complex and of the A/Fd that become over reduced by the feeding of electrons through the Ndh complex and by the low electron drain at the level of plastoquinol peroxidase due to the low concentration of the substrate $H_2O_2$. The increased over-reduction of transporters leads to progressive increases of the production of $O_2\bullet^-$ which, complemented by falling induction of SOD, decreases the ratio of $H_2O_2$ to $O_2\bullet^-$. In addition, not only $O_2\bullet^-$ increases, over-reduced transporters hinder the transfer of electrons from the Chl* of PSII to LET, which facilitates its transition to $^3$Chl* (chloropyll excited at the triplet state) that transfers excitation to $O_2$ enhancing the generation of more $^1O_2$. The generation of $^1O_2$ is further favored because over-reduction decreases CET, which impairs the dissipation as heat of the light energy absorbed to form Chl*. Therefore, when compared with young leaves, changes of enzyme activities increase levels of the ROS $O_2\bullet^-$ and $^1O_2$ in respect to $H_2O_2$ in response to mild stress in early senescent leaves. The changes are auto-catalytically amplified by the functional properties of the photosynthetic machinery, cancel the defense response mediated by $H_2O_2$ against stress and lead to senescence processes mediated by $O_2\bullet^-$ and/or $^1O_2$. In contrast to young leaves, in senescent leaves, mild stress shoots uncontrolled increases of $^1O_2$ and $O_2\bullet^-$. Accordingly, Krieger-Liszkay et al. (2015) observed that senescing flag leaves of two varieties of barley show increased levels of $^1O_2$ or $O_2\bullet^-$.

Low levels of SOD are associated with senescence in animals (Orr and Sohal 1994) and plants (Bowler et al. 1992; Casano et al. 1994, 1999; Abarca et al. 2001b), and a question on the control of leaf cell death by ROS refers to the failure to express the nuclear genes for chloroplastic SODs at advanced developmental stages of the leaf. Genes for chloroplastic SODs are commonly expressed during chloroplast development and (similarly to genes encoding proteins of enzymes of Benson-Calvin cycle, light-harvesting complexes, photosystems and photosynthetic electron transport) their expression is stimulated by light, which agrees with a functional role of chloroplastic SODs closely associated to photosynthesis. Accordingly, there are several TATA and AGATAA sequences up-stream of SOD genes that are involved in the binding of the transcriptional complex and promoter sequences specific for photosynthetic genes, as GT-elements (GGTTAA) involved in the expression of genes for the photosynthetic machinery (Zhou 1999). In addition, the expression of SOD genes increases when leaves are subjected to photo oxidative stress. That is, when excess of light or inhibitors of the photosynthetic electron transport increases the formation of $O_2\bullet^-$ due to transfer of electrons from over-reduced A/Fd to oxygen. This recalls the protective function of SOD against stress by scavenging $O_2\bullet^-$. Less is known about factors responsible for the failure to induce SOD in senescent leaves. One probable clue may be provided by the comparison of the up-stream sequences of the Cu/Zn SOD genes in Arabidopsis ecotypes Ler and Cvi. The Cvi ecotype shows delayed senescence (Luquez et al. 2006) and a distinctive Cu/Zn SOD that remains inducible by photo-oxidative stress at advanced plant age (Abarca et al. 2001b). In respect to the standard Ler ecotype, upstream of Cu/Zn SOD gene, Cvi has a 20-base deletion immediately down-stream of one AGATAA and two TATA box motifs (position −257 to −238) (Sabater and Martín 2013a,b). The sequence deleted in Cvi includes the AACTAA motif recognized by some MYB transcriptional factors that, upon binding it inhibits the transcription of downstream genes (Jin et al. 2000). It is tempting to suggest that in the Ler ecotype of Arabidopsis, and in general in normal plants, one MYB factor specifically accumulates at appropriate advanced development of the leaf, binds to the AACTAA motif and inhibits the transcription of the Cu/Zn SOD gene. Future progress in understanding the decline of SOD, which leads to leaf senescence, is hindered by the complex network of signals (that include a high number of MYB factors) controlling gene expression during plant development.

The other key feature in the control of programmed cell death of leaves by ROS is the autocatalytic increase of the Ndh complex and PX despite the falling induction of chloroplastic SOD. Some investigations with the Ndh complex provide insights to this question and the signal cascade down $^1O_2$ in programmed cell death.

## Signal Cascade Downstream ROS

$H_2O_2$ (frequently generated in chloroplasts) increases in the cytoplasm in the systemic and hyper-sensible defense responses (Levine 1999; Van Breusegem et al. 2001; Overmyer et al. 2003) and acts as a signal of transduction cascades, probably involving mitogen-activated protein kinases, which induce in the nucleus the expression of genes protecting against different stresses (Piñas and Strand 2008). Cytosolic and chloroplastic 2-cysteine peroxiredoxins can further modulate the level of $H_2O_2$ in response to stress (Dietz 2003; Muthuramalingam et al. 2009), but the protein that firstly interacts with $H_2O_2$ in network signaling has not yet been identified.

High level of $H_2O_2$ increases the expression of the *ndh* genes encoding the thylakoid Ndh complex in an action depending on protein kinases (Casano et al. 2001; Lascano et al. 2003). The increased activity of the Ndh complex may explain the stimulation of CET by $H_2O_2$ providing the link between environmental stress, metabolism, and redox regulation of CET in higher plants (Stranda et al. 2015). However, the increase of Ndh during senescence is not mediated by increases of $H_2O_2$ and, probably, other ROS-derived signals induce processes related to programmed cell death into and outside of chloroplasts. As pointed above, increase of PX and decrease of SOD suggest that the production of $H_2O_2$ decreases early in senescence and, accordingly, bundle sheaths in the maize do not accumulate $H_2O_2$ during programmed cell death (Huang and Braun 2010). Accordingly, an increase of the ratio of $^1O_2$ plus $O_2•^-$ to $H_2O_2$ marks the transition from a response of defense against stress to a response of programmed cell death in leaves (Sabater and Martín 2013a,b).

Low half-life and permeation through membranes suggest that the immediate effect of $^1O_2$ and $O_2•^-$ in the signal cascade that shoots programmed cell death must occur within the chloroplasts. $O_2•^-$ could initiate a signal chain by oxidation of cysteine, free and forming part of proteins, but successive steps (if any) leading to programmed cell death are unknown. Alternatively, unspecific oxidation by $O_2•^-$ damages components of the photosynthetic machinery, increasing the production of $^1O_2$ by transfer of excitation from $^3$Chl* to oxygen because Chl* can barely transfer electrons to damaged transporters. Therefore, direct or non-directly, $^1O_2$ would be the key signal that leads the cell to death. One of the best-known effects of $^1O_2$ is that it attacks polyunsaturated fatty acids. Several evidences suggest that this may be the distinctive starting point leading to specific expression of genes for programmed cell death. When the linoleic acid components of membrane lipids react with $^1O_2$ they form 13-hydroperoxy linoleic acid, which is transformed in chloroplasts to the oxylipin (9S,13S)-12-oxo-phytodienoic acid. Oxylipins exit chloroplasts and (9S,13S)-12-oxo-phytodienoic acid is transformed to jasmonic acid in peroxisomes, which is methylated to methyl jasmonate in cytosol (Creelman and Mulpuri 2002; Wasternack 2007).

Jasmonic acid and methyl jasmonate are known mediators of the responses to stress (Laloi and Havaux 2015), where there is over-production of $^1O_2$ (Wagner et al. 2004), and stimulate the expression of the senescence-associated genes (SAG) (Creelman and Mulpuri 2002), among them probably MYB transcription factors (Chen et al. 2002; Buchanan Wollaston et al. 2005). In addition, jasmonic acid and methyl jasmonate increase the levels of the thylakoid Ndh complex (Cuello et al. 1995) and of chloroplast lipoxygenase (Bachmann et al. 2002). These two actions seem key to complete a first turn of the spiral rise of $^1O_2$ and $O_2•^-$ because they are not accompanied by a rise of the concentration of $H_2O_2$ due to the low SOD activity. The progressively misbalanced levels of the different ROS probably determine the programmed cell death (Sabater and Martín 2013a,b). In effect, as explained above, the increase of the thylakoid Ndh complex raises the reduction level of electron transporters and the rate of the Mehler reaction, which, without increase of SOD, further raises the level of $O_2{\bullet}^-$. In addition, lipoxygenase catalyzes the oxidation of free linoleic acid with non-excited molecular oxygen ($O_2$) to form 13-hydroperoxy linoleic acid (Schaller et al. 2005) later transformed to jasmonic acid and methyl jasmonate as indicated. This determines an autocatalytic increase of the $^1O_2$ and jasmonates (mainly jasmonic acid and methyl jasmonate) signals that lead to the death of the cell. Significantly, jasmonic acid increases the production of $^1O_2$ (Guo et al. 2010) and accelerates senescence (Wasternack 2007). The possibility exists that the gene encoding a MYB factor involved in the negative control of chloroplastic SOD expression is one of the SAG jasmonate-mediated induced by $^1O_2$. If true, a suggestive model would emerge where a few signals and transcription factors initiate the autocatalytic path mediated by ROS and lead to leaf cell death.

The increase of the thylakoid Ndh complex is another key aspect of the autocatalytic travel to cell death. Accordingly, leaves of transgenic tobacco defective in the Ndh complex show delayed senescence when compared with normal plants containing full functional Ndh complex (Zapata et al. 2005) and the expression of *ndh* genes and the level of the Ndh complex increases during leaf senescence and fruit ripening (Martín et al. 1996; Casano et al. 1999, 2000; Lascano et al. 2003; Nashilevitz et al. 2010; Nilo et al. 2012; Serrot et al. 2012). In this respect, the longevity (more than 30 years) of the needles of many conifers may be related to the absence of *ndh* genes in their plastid DNA sequenced so far (Wu et al. 2011).

## Conclusion

Many questions remain on the molecular mechanisms through which ROS controls cell death. The identification of the transcription factors involved is one aspect of the complexity of the network of signals controlling cell development through the expression of specific genes and by affecting the activity of enzymes and protein in general. Several hormonal and environmental factors that affect leaf senescence may be involved in the loss of sensitivity of the expression of chloroplast SOD genes to photo-oxidative stress and in the successive autocatalytic events. Further investigations must solve the fragmentary knowledge of the signal factors and their interactions, where ROS are important but not the only players. However, as deduced from experimental results and plausible hypothesis, the actual knowledge of the control of leaf senescence by ROS provides valuable guides for investigation ahead.

A plant invests most of its assets (usually nutrients) to do efficient conversion of light energy into chemical energy of biomass in photosynthesis. The intensity of light and other factors, such as water availability and temperature, strongly varies with time. Therefore, the selection of the photosynthetic machinery traits during evolution have confronted a trade-off between the specialization for a narrow variety of light-environments to optimize photosynthesis or the use of a wide variety of light-environments where more light is available for energy conversion although with low yield. The results of trade-off issues greatly vary at the level of leaves among different plants, but they all lay between defenses against stress conditions to perform photosynthesis and die to recover nutrients in seeds (monocarpic senescence), in upper new leaves (senescence of basal shadowed leaves) or for new leaves in the following year (autumnal senescence). Not surprising, at least for the involvement of ROS in defense and death, programmed cell death uses a subtle modification of the mechanism through which ROS direct the response of defense against stress.

Most abiotic stresses provoke photo-oxidative stress in leaves in which over-reduction of photosynthetic electron transporters generates ROS. Leaves have enzyme and non-enzyme systems to diminish the production of ROS and scavenge them. The different ROS, by still not completely known mechanisms but where $H_2O_2$ seems to play a predominant role, act as signals that activate feed-back responses to alleviate the stress and decrease the level of ROS. Increased activities of chlororespiratory electron transport are one of the protective responses. Chlororespiration transforms the highly reactive ROS $O_2\bullet^-$ in the mild ROS $H_2O_2$ and, maintaining active CET, decreases the production of the also highly reactive ROS $^1O_2$. The thylakoid Ndh complex completes the chlororespiratory chain to maintain active CET in the transition to non-stressing conditions. This is particularly important for very rapid transitions (in less than one seconds) as in sun fleck, where the Ndh complex is required for rapid recovery of photosynthesis rates (Martín et al. 2015). Coordinated induction of the Ndh complex, SOD and PX ensures low levels and appropriate ratios of ROS to alleviate damages and induce defense responses. For still poorly known mechanisms, at specific developmental stages and age of the leaves, coordination broke down by failing the induction of SOD. Then, mild stress is sufficient to start a response where the ratio $O_2{\bullet}^-$ plus $^1O_2$ to $H_2O_2$ feed-forward increases, preventing defense responses and leading to death through mechanisms involving lipoxygenase, jasmonates and different SAG (Sabater and Martín 2013a,b; Laloi and Havaux 2015). In other words: the program of cell leaf senescence enters into action with small changes of a program (defense against stress) designed for the normal functioning of the cell under variable environments.

## REFERENCES

Abarca, D., Martín, M. & Sabater, B. (2001a). Differential leaf stress responses in young and senescent plants. *Physiologia Plantarum*, *113*, 409-415.

Abarca, D., Roldán, M., Martín, M. & Sabater, B. (2001b). *Arabidopsis thaliana* ecotype Cvi shows an increased tolerance to photo-oxidative stress and contains a new chloroplastic copper/zinc superoxide dismutase isoenzyme. *Journal of Experimental Botany*, *52*, 1417-1425.

Apel, K., Hirt, H. (2004). Reactive oxygen species: metabolism, oxidative stress, and signal transduction. *Annual Review of Plant Biology*, *55*, 373-379.

Bachmann, A., Hause, B., Maucher, H., Garbe, E., Vörös, K., Weicher, H., Wasternack, C. & Feussner, I. (2002). Jasmonate-induced lipid peroxidation in barley leaves initiated by distinct 13-LOX forms of chloroplasts. *Biological Chemistry*, 383, 1645-1657.

Bowler, C., Van Montagu, M. & Inzé, D. (1992). Superoxide dismutase and stress tolerance. *Annual Review of Plant Physiology and Plant Molecular Biology*, *43*, 83-116.

Buchanan-Wollaston, V., Page, T., Harrison, E., Breeze, E., Pyung, OL., Nam, H. G., Lin, J. F., Wu, S. H., Swidzinski, J., Ishizaki, K. & Leaver, C. (2005). Comparative transcriptome analysis reveals significant differences in gene expression and signalling pathways between developmental and dark/starvation-induced senescence in *Arabidopsis*. *The Plant Journal*, *42*, 567-585.

Casano, L. M., Martín, M. & Sabater, B. (1994). Sensitivity of superoxide dismutase transcript levels and activities to oxidative stress is lower in mature-senescent than in young barley leaves. *Plant Physiology*, *106*, 1033-1039.

Casano, L. M., Martín, M. & Sabater, B. (2001). Hydrogen peroxide mediates the induction of chloroplastic Ndh complex under photooxidative stress in barley. *Plant Physiology*, *125*, 1450-1458.

Casano, L. M., Martín, M., Zapata, J. M. & Sabater, B. (1999). Leaf age- and paraquat-dependent effects on the levels of enzymes protecting against photooxidative stress. *Plant Science*, *149*, 13-22.

Casano, L. M., Zapata, J. M., Martín, M. & Sabater, B. (2000). Chlororespiration and poising of cyclic electron transport: plastoquinone as electron transporter between thylakoid NADH dehydrogenase and peroxidase. *Journal of Biological Chemistry*, *275*, 942-948.

Chen, S., Yin, C., Qiang, S., Zhou, F. & Dai, X. (2010). Chloroplastic oxidative burst induced by tenuazonic acid, a natural photosynthesis inhibitor, trigger cell necrosis in *Eupatorium adenophorum* Spreng. *Biochimica et Biophysica Acta*, *1797*, 391-405.

Chen, W., Provart, N. J., Glazebrook, J., Katagiri, F., Chang, H. S., Eulgem, T., Mauch, F., Luan, S., Zou, G., Whitham, S. A., Budworth, P. R., Tao, Y., Xie, Z., Chen, X., Lam, S., Kreps, J. A., Harper, J. F., Si-Ammour, A., Mauch-Mani, B., Heinlein, M., Kobayashi, K., Hohn, T., Dangl, J. L., Wang, X. & Zhu, T. (2002). Expression profile matrix of Arabidopsis transcription factor genes suggests their putative functions in response to environmental stresses. *The Plant Cell*, *14*, 559-74.

Creelman, R. A. & Mulpuri, R. (2002). The oxylipin pathaway in *Arabidopsis*. The Arabidopsis Book, pp 1-24. http:// www.aspb.org/ publications/arabidopsis/. American Society of Plant Physiologists, Rockville.

Cuello, J., Quiles, M. J., Rosauro, J. & Sabater, B. (1995). Effects of growth regulators and light on chloroplast NAD(P)H dehydrogenase activities of senescent barley leaves. *Plant Growth and Regulation*, *17*, 225-232.

Dalling, M. J. (1985). The physiological basis of nitrogen redistribution during grain filling in cereals. In: Harper JE, Schrader LE, Howell RW (eds) Explotation of Physiological and Genetic Variability to Enhance Crop Productivity, pp 55-71. American Society of Plant Physiologists, Rockville.

Dietz, K. J. (2003). Plant Peroxiredoxins. *Annual Review of Plant Biology*, *54*, 93-107.

Doyle, S. M., Diamond, M. & McCabe, P. F. (2010). Chloroplast and reactive oxygen species involvement in apoptotic-like programmed cell death in Arabidopsis suspension cultures. *Journal of Experimental Botany*, *61*, 473-482.

Eskling, M., Emanuelsson, A. & Akerlund, H. E. (2001). Enzyme and mechanisms for violaxanthin-zeaxanthin conversions. In: Aro EM and Andersson B (eds) Regulation of Photosynthesis, pp 433-452. Kluwer, Dordrecht.

Feller, U. & Fischer, A. (1994). Nitrogen metabolism in senescing leaves. *Critical Reviews in Plant Science*, *13*, 241-273.

Gechev, T. S., Van Breusegem, F., Stone, J. M., Denev, I. & Laloi, C. (2006). Reactive oxygen species as signals that modulate plant stress responses and programmed cell death. *BioEssays*, *28*, 1091-1101.

Guo, X. X., Yan, X. Q., Yang, R. Y. & Zeng, Q. P. (2010). Salicylic acid and methyl jasmonate but not Rose Bengal enhance artemisinin production through invoking burst of endogen singlet oxygen. *Plant Science*, *178*, 390-397.

Huang, M. & Braun, D. M. (2010). Genetic analyses of cell death in maize (*Zea mays*, Poaceae) leaves reveal a distinct pathway operating in the *camouflage1* mutant. *American Journal of Botany*, *97*, 357-364.

Jin, H., Cominelli, E., Bailey, P., Parr, A., Mehrtens, F., Jones, J., Tonelli, C., Weisshaar, B. & Martin, C. (2000). Transcriptional repression by AtMYB4 controls production of UV-protecting sunscreens in *Arabidopsis*. *EMBO Journal*, *22*, 6150-6161.

Joët, T., Cournac, L., Peltier, G. & Havaux, M., (2002). Cyclic electron flow around photosystem I in C3 plants. *In vivo* control by the redox state of chloroplasts and involvement of the NADH-dehydrogenase complex. *Plant Physiology*, *128*, 760-769.

Krieger-Liszkay, A., Trösch, M. & Krupinska, K. (2015). Generation of reactive oxygen species in thylakoids from senescing flag leaves of the barley varieties Lomerit and Carina. *Planta*, *241*, 1497-508.

Kurepa, J., Hérouart, D., Van Montagu, M. & Inzé, D. (1997). Differential expression of CuZn- and Fe-superoxide dismutase genes of tobacco during development, oxidative stress and hormonal treatments. *Plant and Cell Physiology*, *38*, 463-470.

Laloi, C. & Havaux, M. (2015). Key players of singlet oxygen-induced cell death in plants. *Frontiers in Plant Science*, *6*, art. 39.

Lascano, H. R., Casano, L. M., Martín, M. & Sabater, B. (2003). The activity of the chloroplastic Ndh complex is regulated by phosphorylation of the NDH-F subunit. *Plant Physiology*, *132*, 256-262.

Levine, A. (1999). Oxidative stress as a regulator of environmental responses in plants. In: Lerner HR (ed) Plant Responses to Environmental Stress:

from Phytohormones to Genome Organization, pp 247-264. Marcel Dekker, Inc. New York

Luquez, V. M. C., Sasal, Y., Medrano, M., Martín, M. I., Mujica, M. & Guiamét, J. J. (2006). Quantitative trait loci analysis of leaf and plant longevity in *Arabidopsis thaliana*. *Journal of Experimental Botany*, *57*, 1361-1372.

Martín, M., Casano, L. M. & Sabater, B. (1996). Identification of the product of *ndhA* gene as a thylakoid protein synthesized in response to photooxidative treatment. *Plant and Cell Physiology*, *37*, 293-298.

Martín, M., Casano, L. M., Zapata, J. M., Guéra, A., del Campo, E. M., Schmitz-Linneweber, C., Maier, R. M. & Sabater, B. (2004). Role of thylakoid Ndh complex and peroxidase in the protection against photo-oxidative stress: fluorescence and enzyme activities in wild-type and *ndhF*-deficient tobacco. *Physiologia Plantarum*, *122*, 443-452.

Martín, M., Marín, D., Serrot, P. H. & Sabater, B. (2015). The rise of the photosynthetic rate when light intensity increases is delayed in *ndh* gene-defective tobacco at high but not at low $CO_2$ concentrations. *Frontiers in Plant Science*, *6*, art. 34.

Morita, K. (1980) Release of nitrogen from chloroplasts during leaf senescence in rice (Oryza sativa L.). *Annals of Botany*, *46*, 297-302.

Mubarakshina, M. M., Ivanov, B. N., Naydov, I. A., Hillier, W., Badger, M. R. & Krieger-Liszkay, A. (2010). Production and diffusion of chloroplastic $H_2O_2$ and its implication to signaling. *Journal of Experimental Botany*, *61*, 3577-3587.

Muthuramalingam, M., Seidel, T., Laxa, M., Nunes de Miranda, S. M., Gärtner, F., Ströher, E., Kandlbinder, A. & Dietz, K. J. (2009). Multiple redox and non-redox interactions define 2-Cys peroxiredoxin as a regulatory hub in the chloroplast. *Molecular Plant*, *2*, 1273-1288.

Nashilevitz, S., Melamed-Bessudo, C., Izkovich, Y., Rogachev, I., Osorio, S., Itkin, M., Adato, A., Pankratov, I., Hirschberg, J., Fernie, AR., Wolf, S., Usadel, B., Levy, A. A., Rumeau, D. & Aharoni, A. (2010) An orange ripening mutant links plastid NAD(P)H dehydrogenase complex activity to central and specialized metabolism during tomato fruit maturation. *The Plant Cell*, *22*, 1977-1997.

Nilo, R. P., Campos-Vargas, R. & Orellana, A. (2012) Assessment of *Prunus persica* fruit softening using a proteomics approach. *Journal of Proteomics*, *75*, 1618-1638.

Nixon, P. J. & Rich, P. R. (2007). Chlororespiratory pathways and their physiological significance. In: Wise RR and Hoober JK (eds) The Structure and Function of Plastids, pp 237-251. Springer, Berlin.

Ohe, M., Rapolu, M., Mieda, T., Miyagawa, Y., Yabuta, Y., Yoshimura, K. & Shigeoka, S. (2005). Decline of leaf photooxidative stress tolerance with age in tobacco. *Plant Science*, *168*, 1487-1493.

Orr, W. C. & Sohal, R. S. (1994). Extension of life-span by overexpression of superoxide dismutase and catalase in *Drosophila melanogaster*. *Science*, *263*, 1128-1130.

Overmyer, K., Brosché, M. & Kangasjärvi, J. (2003). Reactive oxygen species and hormonal control of cell death. *TRENDS in Plant Science*, *8*, 335-342.

Paredes, M. & Quiles, M. J. (2013). Stimulation of chlororespiration by drought under heat and high illumination in *Rosa meillandina*. *Journal of Plant Physiology*, *170*, 165-171.

Piñas, A. & Strand, Å. (2008). Retrograde signaling and plant stress: plastid signals initiate cellular stress responses. *Current Opinion in Plant Biology*, *11*, 509-513.

Prochazkova, D., Sairam, R. K., Srivastava, G. C. & Singh, D. V. (2001) Oxidative stress and antioxidant activity as the basis of senescence in maize. *Plant Science*, *161*, 765-771.

Sabater, B. & Martín, M. (2013a). Chloroplast control of leaf senescence, in Plant development in leaves during growth and senescence. Advances in photosynthesis and respiration. Vol. 36, eds. Biswal B, Krupinska K, Biswal UC, pp 529-550 (Springer, Netherlands).

Sabater, B. & Martín, M. (2013b). Hypothesis: increase of the ratio singlet oxygen plus superoxide radical to hydrogen peroxide changes stress defense response to programmed leaf death. *Frontiers in Plant Science*, *4*, art. 479.

Schaller, F., Schalller, A. & Stintzi, A. (2005). Biosynthesis and metabolism of jasmonates. *Journal of Plant Growth Regulation*, *23*, 179-199.

Segura, M. V. & Quiles, M. J. (2015). Involvement of chlororespiration in chilling stress in the tropical species *Spathiphyllum wallisii*. *Plant Cell and Environment*, *38*, 525-533.

Serrot, P. H., Sabater, B. & Martín, M. (2008). Expression of the *ndhCKJ* operon of barley and editing at the $13^{th}$ base of the mRNA of the *ndhC* gene. *Biologia Plantarum*, *52*, 347-350.

Serrot, P. H., Sabater, B. & Martín M (2012). Activity, polypeptide and gene identification of thylakoid Ndh complex in trees: potential physiological relevance of fluorescence assays. *Physiologia Plantarum*, *146*, 110-120.

Stranda, D. D., Livingstonc, A. K., Satoh-Cruza, M., Froehlicha, J. E., Maurinof, V. G. & Kramer, D. M. (2015). Activation of cyclic electron flow by hydrogen peroxide *in vivo*. *Proceedings of the National Academic of Sciences USA*, *112*, 5539–5544.

Van Breusegem, F., Vranová, E., Dat, J. F. & Inzé, D. (2001). The role of active oxygen species in plant signal transduction. *Plant Science*, *161*, 405-414.

Van Doorn, W. G. & Yoshimoto, K. (2010). Role of chloroplasts and other plastids in ageing and death of plants and animals: A tale of Vishnu and Shiva. *Ageing Research Review*, *9*, 117-130.

Wagner, D., Przybyla, D., op den Camp, R., Kim, C., Landgraf, F., Lee, K. P., Würsch, M., Laloi, C., Nater, M., Hideg, E. & Apel, K. (2004). The genetic basis of singlet oxygen-induced stress responses of *Arabidopsis thaliana*. *Science*, *306*, 1183-1185.

Wasternack, C. (2007). Jasmonates: an update on biosynthesis, signal transduction and action in plant stress response, growth and development. *Annals of Botany*, *100*, 681-697.

Wiciarz, M., Gubernator, B., Kruk, J. & Niewiadomska, E. (2015). Enhanced chloroplastic generation of $H_2O_2$ in stress-resistant *Thellungiella salsuginea* in comparison to *Arabidopsis thaliana*. *Physiologia Plantarum*, *153*, 467-476.

Wu, C. S., Wang, Y. N., Hsu, C. Y., Lin, C. P. & Chaw, S. M. (2011). Loss of different inverted repeat copies from the chloroplast genomes of pinaceae and cupressophytes and influence of heterotachy on the evaluation of gymnosperm phylogeny. *Genome Biology and Evolution*, *3*, 1284-1295.

Zapata, J. M., Guéra, A., Esteban-Carrasco, A., Martín, M. & Sabater, B. (2005). Chloroplasts regulate leaf senescence: delayed senescence in transgenic *ndhF*-defective tobacco. *Cell Death and Differentiation*, *12*, 1277-1284.

Zapata, J. M., Sabater, B. & Martín, M. (1998). Identification of a thylakoid peroxidase which oxydized hydroquinone. *Phytochemistry*, *48*, 1119-1123.

Zhou, D. X. (1999). Regulatory mechanism of plant gene transcription by GT-elements and GT-factors. *Trends in Plant Science*, *4*, 210-214.

# INDEX

## #



## A



## B



## C

## D

## E



## F



## G



## H

## I



## J



## K



## L



## M

## N



## O



## P

## R